T0219994

Data Science Fundamentals for Python and MongoDB

David Paper

Apress®

Data Science Fundamentals for Python and MongoDB

David Paper
Logan, Utah, USA

ISBN-13 (pbk): 978-1-4842-3596-6   ISBN-13 (electronic): 978-1-4842-3597-3
https://doi.org/10.1007/978-1-4842-3597-3

Library of Congress Control Number: 2018941864

Copyright © 2018 by David Paper

This work is subject to copyright. All rights are reserved by the Publisher, whether the whole or part of the material is concerned, specifically the rights of translation, reprinting, reuse of illustrations, recitation, broadcasting, reproduction on microfilms or in any other physical way, and transmission or information storage and retrieval, electronic adaptation, computer software, or by similar or dissimilar methodology now known or hereafter developed.

Trademarked names, logos, and images may appear in this book. Rather than use a trademark symbol with every occurrence of a trademarked name, logo, or image, we use the names, logos, and images only in an editorial fashion and to the benefit of the trademark owner, with no intention of infringement of the trademark.

The use in this publication of trade names, trademarks, service marks, and similar terms, even if they are not identified as such, is not to be taken as an expression of opinion as to whether or not they are subject to proprietary rights.

While the advice and information in this book are believed to be true and accurate at the date of publication, neither the authors nor the editors nor the publisher can accept any legal responsibility for any errors or omissions that may be made. The publisher makes no warranty, express or implied, with respect to the material contained herein.

Managing Director, Apress Media LLC: Welmoed Spahr
Acquisitions Editor: Jonathan Gennick
Development Editor: Laura Berendson
Coordinating Editor: Jill Balzano

Cover designed by eStudioCalamar

Cover image designed by Freepik (www.freepik.com)

Distributed to the book trade worldwide by Springer Science+Business Media New York, 233 Spring Street, 6th Floor, New York, NY 10013. Phone 1-800-SPRINGER, fax (201) 348-4505, e-mail orders-ny@springer-sbm.com, or visit www.springeronline.com. Apress Media, LLC is a California LLC and the sole member (owner) is Springer Science + Business Media Finance Inc (SSBM Finance Inc). SSBM Finance Inc is a **Delaware** corporation.

For information on translations, please e-mail rights@apress.com, or visit http://www.apress.com/rights-permissions.

Apress titles may be purchased in bulk for academic, corporate, or promotional use. eBook versions and licenses are also available for most titles. For more information, reference our Print and eBook Bulk Sales web page at http://www.apress.com/bulk-sales.

Any source code or other supplementary material referenced by the author in this book is available to readers on GitHub via the book's product page, located at www.apress.com/9781484235966. For more detailed information, please visit http://www.apress.com/source-code.

Printed on acid-free paper

To Lady, Sam, Bruce, Malik, John, Moonshadow, and Moonbeam whose support and love is and always has been unconditional. To the Apress staff for all of your support and hard work in making this project happen. Finally, a special shout-out to Jonathan for finding me on Amazon, Jill for putting up with a compulsive author, and Mark for a thourough and constructive technical review.

Table of Contents

About the Author

 **David Paper** is a full professor at Utah State University in the Management Information Systems department. His book *Web Programming for Business: PHP Object-Oriented Programming with Oracle* was published in 2015 by Routledge. He also has over 70 publications in refereed journals such as *Organizational Research Methods, Communications of the ACM, Information & Management, Information Resource Management Journal, Communications of the AIS, Journal of Information Technology Case and Application Research*, and *Long Range Planning*. He has also served on several editorial boards in various capacities, including associate editor. Besides growing up in family businesses, Dr. Paper has worked for Texas Instruments, DLS, Inc., and the Phoenix Small Business Administration. He has performed IS consulting work for IBM, AT&T, Octel, Utah Department of Transportation, and the Space Dynamics Laboratory. Dr. Paper's teaching and research interests include data science, machine learning, process reengineering, object-oriented programming, electronic customer relationship management, change management, e-commerce, and enterprise integration.

About the Technical Reviewer

Mark Furman, MBA is a systems engineer, author, teacher, and entrepreneur. For the last 16 years he has worked in the Information Technology field, with a focus on Linux-based systems and programming in Python, working for a range of companies including Host Gator, Interland, Suntrust Bank, AT&T, and Winn-Dixie. Currently he has been focusing his career on the maker movement and has launched Tech Forge (techforge.org), which will focus on helping people start a makerspace and help sustain current spaces. He holds a Master of Business Administration from Ohio University. You can follow him on Twitter @mfurman.

Acknowledgments

My entrée into data analysis started by exploring *Python for Data Analysis* by Wes McKinney, which I highly recommend to everyone. My entrée into data science started by exploring *Data Science from Scratch* by Joel Grus. Joel's book may not be for the faint of heart, but it is definitely a challenge that I am glad that I accepted! Finally, I thank all of the contributors to *stackoverflow*, whose programming solutions are indispensable.

CHAPTER 1

Introduction

Data science is an interdisciplinary field encompassing scientific methods, processes, and systems to extract knowledge or insights from data in various forms, either structured or unstructured. It draws principles from mathematics, statistics, information science, computer science, machine learning, visualization, data mining, and predictive analytics. However, it is fundamentally grounded in mathematics.

This book explains and applies the fundamentals of data science crucial for technical professionals such as DBAs and developers who are making career moves toward practicing data science. It is an example-driven book providing complete Python coding examples to complement and clarify data science concepts, and enrich the learning experience. Coding examples include visualizations whenever appropriate. The book is a necessary precursor to applying and implementing machine learning algorithms, because it introduces the reader to foundational principles of the science of data.

The book is self-contained. All the math, statistics, stochastic, and programming skills required to master the content are covered in the book. In-depth knowledge of object-oriented programming isn't required, because working and complete examples are provided and explained. The examples are in-depth and complex when necessary to ensure the acquisition of appropriate data science acumen. The book helps you to build the foundational skills necessary to work with and understand complex data science algorithms.

© David Paper 2018
D. Paper, *Data Science Fundamentals for Python and MongoDB*,
https://doi.org/10.1007/978-1-4842-3597-3_1

Data Science Fundamentals by Example is an excellent starting point for those interested in pursuing a career in data science. Like any science, the fundamentals of data science are prerequisite to competency. Without proficiency in mathematics, statistics, data manipulation, and coding, the path to success is "rocky" at best. The coding examples in this book are concise, accurate, and complete, and perfectly complement the data science concepts introduced.

The book is organized into six chapters. Chapter 1 introduces the programming fundamentals with "Python" necessary to work with, transform, and process data for data science applications. Chapter 2 introduces Monte Carlo simulation for decision making, and data distributions for statistical processing. Chapter 3 introduces linear algebra applied with vectors and matrices. Chapter 4 introduces the gradient descent algorithm that minimizes (or maximizes) functions, which is very important because most data science problems are optimization problems. Chapter 5 focuses on munging, cleaning, and transforming data for solving data science problems. Chapter 6 focusing on exploring data by dimensionality reduction, web scraping, and working with large data sets efficiently.

Python programming code for all coding examples and data files are available for viewing and download through Apress at `www.apress.com/9781484235966`. Specific linking instructions are included on the copyright pages of the book.

To install a Python module, pip is the preferred installer program. So, to install the matplotlib module from an Anaconda prompt: pip install matplotlib. Anaconda is a widely popular open source distribution of Python (and R) for large-scale data processing, predictive analytics, and scientific computing that simplifies package management and deployment. I have worked with other distributions with unsatisfactory results, so I highly recommend Anaconda.

Python Fundamentals

Python has several features that make it well suited for learning and doing data science. It's free, relatively simple to code, easy to understand, and has many useful libraries to facilitate data science problem solving. It also allows quick prototyping of virtually any data science scenario and demonstration of data science concepts in a clear, easy to understand manner.

The goal of this chapter is *not* to teach Python as a whole, but present, explain, and clarify fundamental features of the language (such as logic, data structures, and libraries) that help prototype, apply, and/or solve data science problems.

Python fundamentals are covered with a wide spectrum of activities with associated coding examples as follows:

1. functions and strings

2. lists, tuples, and dictionaries

3. reading and writing data

4. list comprehension

5. generators

6. data randomization

7. MongoDB and JSON

8. visualization

Functions and Strings

Python functions are first-class functions, which means they can be used as parameters, a return value, assigned to variable, and stored in data structures. Simply, functions work like a typical variable. Functions can be either custom or built-in. Custom are created by the programmer, while built-in are part of the language. Strings are very popular types enclosed in either single or double quotes.

The following code example defines custom functions and uses built-in ones:

```python
def num_to_str(n):
    return str(n)

def str_to_int(s):
    return int(s)

def str_to_float(f):
    return float(f)

if __name__ == "__main__":
    # hash symbol allows single-line comments
    '''
    triple quotes allow multi-line comments
    '''
    float_num = 999.01
    int_num = 87
    float_str = '23.09'
    int_str = '19'
    string = 'how now brown cow'
    s_float = num_to_str(float_num)
    s_int = num_to_str(int_num)
    i_str = str_to_int(int_str)
    f_str = str_to_float(float_str)
    print (s_float, 'is', type(s_float))
    print (s_int, 'is', type(s_int))
    print (f_str, 'is', type(f_str))
    print (i_str, 'is', type(i_str))
```

```
print ('\nstring', '"' + string + '" has', len(string),
'characters')
str_ls = string.split()
print ('split string:', str_ls)
print ('joined list:', ' '.join(str_ls))
```

Output:

```
999.01 is <class 'str'>
87 is <class 'str'>
23.09 is <class 'float'>
19 is <class 'int'>

string "how now brown cow" has 17 characters
split string: ['how', 'now', 'brown', 'cow']
joined list: how now brown cow
```

A popular coding style is to present library importation and functions first, followed by the main block of code. The code example begins with three custom functions that convert numbers to strings, strings to numbers, and strings to float respectively. Each custom function returns a built-in function to let Python do the conversion. The main block begins with comments. Single-line comments are denoted with the # (hash) symbol. Multiline comments are denoted with three consecutive single quotes. The next five lines assign values to variables. The following four lines convert each variable type to another type. For instance, function num_to_str() converts variable float_num to string type. The next five lines print variables with their associated Python data type. Built-in function type() returns type of given object. The remaining four lines print and manipulate a string variable.

Lists, Tuples, and Dictionaries

Lists are ordered collections with comma-separated values between square brackets. Indices start at 0 (zero). List items need not be of the same type and can be sliced, concatenated, and manipulated in many ways.

The following code example creates a list, manipulates and slices it, creates a new list and adds elements to it from another list, and creates a matrix from two lists:

```python
import numpy as np

if __name__ == "__main__":
    ls = ['orange', 'banana', 10, 'leaf', 77.009, 'tree', 'cat']
    print ('list length:', len(ls), 'items')
    print ('cat count:', ls.count('cat'), ',', 'cat index:',
    ls.index('cat'))
    print ('\nmanipulate list:')
    cat = ls.pop(6)
    print ('cat:', cat, ', list:', ls)
    ls.insert(0, 'cat')
    ls.append(99)
    print (ls)
    ls[7] = '11'
    print (ls)
    ls.pop(1)
    print (ls)
```

```
ls.pop()
print (ls)
print ('\nslice list:')
print ('1st 3 elements:', ls[:3])
print ('last 3 elements:', ls[3:])
print ('start at 2nd to index 5:', ls[1:5])
print ('start 3 from end to end of list:', ls[-3:])
print ('start from 2nd to next to end of list:', ls[1:-1])
print ('\ncreate new list from another list:')
print ('list:', ls)
fruit = ['orange']
more_fruit = ['apple', 'kiwi', 'pear']
fruit.append(more_fruit)
print ('appended:', fruit)
fruit.pop(1)
fruit.extend(more_fruit)
print ('extended:', fruit)
a, b = fruit[2], fruit[1]
print ('slices:', a, b)
print ('\ncreate matrix from two lists:')
matrix = np.array([ls, fruit])
print (matrix)
print ('1st row:', matrix[0])
print ('2nd row:', matrix[1])
```

Output:

```
list length: 7 items
cat count: 1 , cat index: 6

manipulate list:
cat: cat , list: ['orange', 'banana', 10, 'leaf', 77.009, 'tree']
['cat', 'orange', 'banana', 10, 'leaf', 77.009, 'tree', 99]
['cat', 'orange', 'banana', 10, 'leaf', 77.009, 'tree', '11']
['cat', 'banana', 10, 'leaf', 77.009, 'tree', '11']
['cat', 'banana', 10, 'leaf', 77.009, 'tree']

slice list:
1st 3 elements: ['cat', 'banana', 10]
last 3 elements: ['leaf', 77.009, 'tree']
start at 2nd to index 5: ['banana', 10, 'leaf', 77.009]
start 3 from end to end of list: ['leaf', 77.009, 'tree']
start from 2nd to next to end of list: ['banana', 10, 'leaf', 77.009]

create new list from another list:
list: ['cat', 'banana', 10, 'leaf', 77.009, 'tree']
appended: ['orange', ['apple', 'kiwi', 'pear']]
extended: ['orange', 'apple', 'kiwi', 'pear']
slices: kiwi apple

create matrix from two lists:
[['cat', 'banana', 10, 'leaf', 77.009, 'tree']
 ['orange', 'apple', 'kiwi', 'pear']]
1st row: ['cat', 'banana', 10, 'leaf', 77.009, 'tree']
2nd row: ['orange', 'apple', 'kiwi', 'pear']
```

The code example begins by importing NumPy, which is the fundamental package (library, module) for scientific computing. It is useful for linear algebra, which is fundamental to data science. Think of Python libraries as giant classes with many methods. The main block begins by creating list ls, printing its length, number of elements (items), number of cat elements, and index of the cat element. The code continues by manipulating ls. First, the 7th element (index 6) is popped and assigned to variable cat. Remember, list indices start at 0. Function pop() removes cat from ls. Second, cat is added back to ls at the 1st position (index 0) and 99 is appended to the end of the list. Function append() adds an object to the end of a list. Third, string '11' is substituted for the 8th element (index 7). Finally, the 2nd element and the last element are popped from ls. The code continues by slicing ls. First, print the 1st three elements with ls[:3].

Second, print the last three elements with ls[3:]. Third, print starting with the 2nd element to elements with indices up to 5 with ls[1:5]. Fourth, print starting three elements from the end to the end with ls[-3:]. Fifth, print starting from the 2nd element to next to the last element with ls[1:-1]. The code continues by creating a new list from another. First, create fruit with one element. Second append list more_fruit to fruit. Notice that append adds list more_fruit as the 2nd element of fruit, which may not be what you want. So, third, pop 2nd element of fruit and extend more_fruit to fruit. Function extend() unravels a list before it adds it. This way, fruit now has four elements. Fourth, assign 3rd element to a and 2nd element to b and print slices. Python allows assignment of multiple variables on one line, which is very convenient and concise. The code ends by creating a matrix from two lists—ls and fruit—and printing it. A Python matrix is a two-dimensional (2-D) array consisting of rows and columns, where each row is a list.

A tuple is a sequence of immutable Python objects enclosed by parentheses. Unlike lists, tuples cannot be changed. Tuples are convenient with functions that return multiple values.

The following code example creates a tuple, slices it, creates a list, and creates a matrix from tuple and list:

```python
import numpy as np

if __name__ == "__main__":
    tup = ('orange', 'banana', 'grape', 'apple', 'grape')
    print ('tuple length:', len(tup))
    print ('grape count:', tup.count('grape'))
    print ('\nslice tuple:')
    print ('1st 3 elements:', tup[:3])
    print ('last 3 elements', tup[3:])
    print ('start at 2nd to index 5', tup[1:5])
    print ('start 3 from end to end of tuple:', tup[-3:])
```

```
print ('start from 2nd to next to end of tuple:', tup[1:-1])
print ('\ncreate list and create matrix from it and tuple:')
fruit = ['pear', 'grapefruit', 'cantaloupe', 'kiwi', 'plum']
matrix = np.array([tup, fruit])
print (matrix)
```

Output:

```
tuple length: 5
grape count: 2

slice tuple:
1st 3 elements: ('orange', 'banana', 'grape')
last 3 elements ('apple', 'grape')
start at 2nd to index 5 ('banana', 'grape', 'apple', 'grape')
start 3 from end to end of tuple: ('grape', 'apple', 'grape')
start from 2nd to next to end of tuple: ('banana', 'grape', 'apple')

create list and create matrix from it and tuple:
[['orange' 'banana' 'grape' 'apple' 'grape']
 ['pear' 'grapefruit' 'cantaloupe' 'kiwi' 'plum']]
```

The code begins by importing NumPy. The main block begins by creating tuple tup, printing its length, number of elements (items), number of grape elements, and index of grape. The code continues by slicing tup. First, print the 1st three elements with tup[:3]. Second, print the last three elements with tup[3:]. Third, print starting with the 2nd element to elements with indices up to 5 with tup[1:5]. Fourth, print starting three elements from the end to the end with tup[-3:]. Fifth, print starting from the 2nd element to next to the last element with tup[1:-1]. The code continues by creating a new fruit list and creating a matrix from tup and fruit.

A dictionary is an unordered collection of items identified by a key/value pair. It is an extremely important data structure for working with data. The following example is very simple, but the next section presents a more complex example based on a dataset.

The following code example creates a dictionary, deletes an element, adds an element, creates a list of dictionary elements, and traverses the list:

```python
if __name__ == "__main__":
    audio = {'amp':'Linn', 'preamp':'Luxman', 'speakers':'Energy',
             'ic':'Crystal Ultra', 'pc':'JPS', 'power':'Equi-Tech',
             'sp':'Crystal Ultra', 'cdp':'Nagra', 'up':'Esoteric'}
    del audio['up']
    print ('dict "deleted" element;')
    print (audio, '\n')
    print ('dict "added" element;')
    audio['up'] = 'Oppo'
    print (audio, '\n')
    print ('universal player:', audio['up'], '\n')
    dict_ls = [audio]
    video = {'tv':'LG 65C7 OLED', 'stp':'DISH', 'HDMI':'DH Labs',
             'cable' : 'coax'}
    print ('list of dict elements;')
    dict_ls.append(video)
    for i, row in enumerate(dict_ls):
        print ('row', i, ':')
        print (row)
```

Output:

```
dict "deleted" element;
{'amp': 'Linn', 'preamp': 'Luxman', 'speakers': 'Energy', 'ic': 'Crystal Ultra',
'pc': 'JPS', 'power': 'Equi-Tech', 'sp': 'Crystal Ultra', 'cdp': 'Nagra'}

dict "added" element;
{'amp': 'Linn', 'preamp': 'Luxman', 'speakers': 'Energy', 'ic': 'Crystal Ultra',
'pc': 'JPS', 'power': 'Equi-Tech', 'sp': 'Crystal Ultra', 'cdp': 'Nagra', 'up':
'Oppo'}

universal player: Oppo

list of dict elements;
row 0 :
{'amp': 'Linn', 'preamp': 'Luxman', 'speakers': 'Energy', 'ic': 'Crystal Ultra',
'pc': 'JPS', 'power': 'Equi-Tech', 'sp': 'Crystal Ultra', 'cdp': 'Nagra', 'up':
'Oppo'}
row 1 :
{'tv': 'LG 65C7 OLED', 'stp': 'DISH', 'HDMI': 'DH Labs', 'cable': 'coax'}
```

11

The main block begins by creating dictionary audio with several elements. It continues by deleting an element with key up and value Esoteric, and displaying. Next, a new element with key up and element Oppo is added back and displayed. The next part creates a list with dictionary audio, creates dictionary video, and adds the new dictionary to the list. The final part uses a for loop to traverse the dictionary list and display the two dictionaries. A very useful function that can be used with a loop statement is enumerate(). It adds a counter to an iterable. An iterable is an object that can be iterated. Function enumerate() is very useful because a counter is automatically created and incremented, which means less code.

Reading and Writing Data

The ability to read and write data is fundamental to any data science endeavor. All data files are available on the website. The most basic types of data are text and CSV (Comma Separated Values). So, this is where we will start.

The following code example reads a text file and cleans it for processing. It then reads the precleansed text file, saves it as a CSV file, reads the CSV file, converts it to a list of OrderedDict elements, and converts this list to a list of regular dictionary elements.

```
import csv

def read_txt(f):
    with open(f, 'r') as f:
        d = f.readlines()
        return [x.strip() for x in d]

def conv_csv(t, c):
    data = read_txt(t)
    with open(c, 'w', newline='') as csv_file:
```

```python
        writer = csv.writer(csv_file)
        for line in data:
            ls = line.split()
            writer.writerow(ls)

def read_csv(f):
    contents = ''
    with open(f, 'r') as f:
        reader = csv.reader(f)
        return list(reader)

def read_dict(f, h):
    input_file = csv.DictReader(open(f), fieldnames=h)
    return input_file

def od_to_d(od):
    return dict(od)

if __name__ == "__main__":
    f = 'data/names.txt'
    data = read_txt(f)
    print ('text file data sample:')
    for i, row in enumerate(data):
        if i < 3:
            print (row)
    csv_f = 'data/names.csv'
    conv_csv(f, csv_f)
    r_csv = read_csv(csv_f)
    print ('\ntext to csv sample:')
    for i, row in enumerate(r_csv):
        if i < 3:
            print (row)
    headers = ['first', 'last']
```

13

```
r_dict = read_dict(csv_f, headers)
dict_ls = []
print ('\ncsv to ordered dict sample:')
for i, row in enumerate(r_dict):
    r = od_to_d(row)
    dict_ls.append(r)
    if i < 3:
        print (row)
print ('\nlist of dictionary elements sample:')
for i, row in enumerate(dict_ls):
    if i < 3:
        print (row)
```

Output:

```
text file data sample:
Adam Baum
Adam Zapel
Al Bino

text to csv sample:
['Adam', 'Baum']
['Adam', 'Zapel']
['Al', 'Bino']

csv to ordered dict sample:
OrderedDict([('first', 'Adam'), ('last', 'Baum')])
OrderedDict([('first', 'Adam'), ('last', 'Zapel')])
OrderedDict([('first', 'Al'), ('last', 'Bino')])

list of dictionary elements sample:
{'first': 'Adam', 'last': 'Baum'}
{'first': 'Adam', 'last': 'Zapel'}
{'first': 'Al', 'last': 'Bino'}
```

The code begins by importing the csv library, which implements classes to read and write tabular data in CSV format. It continues with five functions. Function read_txt() reads a text (.txt) file and strips (removes) extraneous characters with list comprehension, which is an elegant way to define and create a list in Python. List comprehension is covered later in the next section. Function conv_csv() converts a text to a CSV file and saves it to disk. Function read_csv() reads a CSV file and returns it as a list. Function read_dict() reads a CSV file and returns a list of OrderedDict elements. An OrderedDict is a dictionary subclass that remembers the order in which its contents are added, whereas a regular dictionary doesn't track insertion order. Finally, function od_to_d() converts an OrderedDict element to a regular dictionary element. Working with a regular dictionary element is much more intuitive in my opinion. The main block begins by reading a text file and cleaning it for processing. However, no processing is done with this cleansed file in the code. It is only included in case you want to know how to accomplish this task. The code continues by converting a text file to CSV, which is saved to disk. The CSV file is then read from disk and a few records are displayed. Next, a headers list is created to store keys for a dictionary yet to be created. List dict_ls is created to hold dictionary elements. The code continues by creating an OrderedDict list r_dict. The OrderedDict list is then iterated so that each element can be converted to a regular dictionary element and appended to dict_ls. A few records are displayed during iteration. Finally, dict_ls is iterated and a few records are displayed. I highly recommend that you take some time to familiarize yourself with these data structures, as they are used extensively in data science application.

List Comprehension

List comprehension provides a concise way to create lists. Its logic is enclosed in square brackets that contain an expression followed by a for clause and can be augmented by more for or if clauses.

The read_txt() function in the previous section included the following list comprehension:

```
[x.strip() for x in d]
```

The logic strips extraneous characters from string in iterable d. In this case, d is a list of strings.

The following code example converts miles to kilometers, manipulates pets, and calculates bonuses with list comprehension:

```
if __name__ == "__main__":
    miles = [100, 10, 9.5, 1000, 30]
    kilometers = [x * 1.60934 for x in miles]
    print ('miles to kilometers:')
    for i, row in enumerate(kilometers):
        print ('{:>4} {:>8}{:>8} {:>2}'.
                format(miles[i],'miles is', round(row,2), 'km'))
    print ('\npet:')
    pet = ['cat', 'dog', 'rabbit', 'parrot', 'guinea pig', 'fish']
    print (pet)
    print ('\npets:')
    pets = [x + 's' if x != 'fish' else x for x in pet]
    print (pets)
    subset = [x for x in pets if x != 'fish' and x != 'rabbits'
                and x != 'parrots' and x != 'guinea pigs']
    print ('\nmost common pets:')
    print (subset[1], 'and', subset[0])
    sales = [9000, 20000, 50000, 100000]
    print ('\nbonuses:')
    bonus = [0 if x < 10000 else x * .02 if x >= 10000
    and x <= 20000
                else x * .03 for x in sales]
    print (bonus)
    print ('\nbonus dict:')
    people = ['dave', 'sue', 'al', 'sukki']
    d = {}
    for i, row in enumerate(people):
```

```
        d[row] = bonus[i]
print (d, '\n')
print ('{:<5} {:<5}'.format('emp', 'bonus'))
for k, y in d.items():
    print ('{:<5} {:>6}'.format(k, y))
```

Output:

```
miles to kilometers:
 100 miles is  160.93 km
  10 miles is   16.09 km
 9.5 miles is   15.29 km
1000 miles is 1609.34 km
  30 miles is   48.28 km

pet:
['cat', 'dog', 'rabbit', 'parrot', 'guinea pig', 'fish']

pets:
['cats', 'dogs', 'rabbits', 'parrots', 'guinea pigs', 'fish']

most common pets:
dogs and cats

bonuses:
[0, 400.0, 1500.0, 3000.0]

bonus dict:
{'dave': 0, 'sue': 400.0, 'al': 1500.0, 'sukki': 3000.0}

emp   bonus
dave      0
sue   400.0
al    1500.0
sukki 3000.0
```

The main block begins by creating two lists—miles and kilometers. The kilometers list is created with list comprehension, which multiplies each mile value by 1.60934. At first, list comprehension may seem confusing, but practice makes it easier over time. The main block continues by printing miles and associated kilometers. Function format() provides sophisticated formatting options. Each mile value is ({:>4}) with up to four characters right justified. Each string for miles and kilometers is right justified ({:>8}) with up to eight characters. Finally, each string for km is right justified ({:>2}) with up to two characters. This may seem a bit complicated at first, but it is really quite logical (and elegant) once you get used to it. The main block continues by creating pet and pets lists. The pets list is created with list comprehension, which makes a pet plural if it is not a fish. I advise you to study this list comprehension before you go forward, because they just get more complex. The code continues by creating a subset list with list comprehension, which only includes dogs and cats. The next part creates two lists—sales and bonus. Bonus is created with list comprehension that calculates bonus for each sales value. If sales are less than 10,000, no bonus is paid. If sales are between 10,000 and 20,000 (inclusive), the bonus is 2% of sales. Finally, if sales if greater than 20,000, the bonus is 3% of sales. At first I was confused with this list comprehension but it makes sense to me now. So, try some of your own and you will get the gist of it. The final part creates a people list to associate with each sales value, continues by creating a dictionary to hold bonus for each person, and ends by iterating dictionary elements. The formatting is quite elegant. The header left justifies emp and bonus properly. Each item is formatted so that the person is left justified with up to five characters ({:<5}) and the bonus is right justified with up to six characters ({:>6}).

Generators

A generator is a special type of iterator, but much faster because values are only produced as needed. This process is known as lazy (or deferred) evaluation. Typical iterators are much slower because they are fully built into memory. While regular functions return values, generators yield them. The best way to traverse and access values from a generator is to use a loop. Finally, a list comprehension can be converted to a generator by replacing square brackets with parentheses.

The following code example reads a CSV file and creates a list of OrderedDict elements. It then converts the list elements into regular dictionary elements. The code continues by simulating times for list comprehension, generator comprehension, and generators. During simulation, a list of times for each is created. Simulation is the imitation of a real-world process or system over time, and it is used extensively in data science.

```python
import csv, time, numpy as np

def read_dict(f, h):
    input_file = csv.DictReader(open(f), fieldnames=h)
    return (input_file)

def conv_reg_dict(d):
    return [dict(x) for x in d]

def sim_times(d, n):
    i = 0
    lsd, lsgc = [], []
    while i < n:
        start = time.clock()
        [x for x in d]
        time_d = time.clock() - start
        lsd.append(time_d)
        start = time.clock()
        (x for x in d)
        time_gc = time.clock() - start
        lsgc.append(time_gc)
        i += 1
    return (lsd, lsgc)
```

```python
def gen(d):
    yield (x for x in d)

def sim_gen(d, n):
    i = 0
    lsg = []
    generator = gen(d)
    while i < n:
        start = time.clock()
        for row in generator:
            None
        time_g = time.clock() - start
        lsg.append(time_g)
        i += 1
        generator = gen(d)
    return lsg

def avg_ls(ls):
    return np.mean(ls)

if __name__ == '__main__':
    f = 'data/names.csv'
    headers = ['first', 'last']
    r_dict = read_dict(f, headers)
    dict_ls = conv_reg_dict(r_dict)
    n = 1000
    ls_times, gc_times = sim_times(dict_ls, n)
    g_times = sim_gen(dict_ls, n)
    avg_ls = np.mean(ls_times)
    avg_gc = np.mean(gc_times)
    avg_g = np.mean(g_times)
    gc_ls = round((avg_ls / avg_gc), 2)
    g_ls = round((avg_ls / avg_g), 2)
```

```
print ('generator comprehension:')
print (gc_ls, 'times faster than list comprehension\n')
print ('generator:')
print (g_ls, 'times faster than list comprehension')
```

Output:

```
generator comprehension:
9.46 times faster than list comprehension

generator:
9.66 times faster than list comprehension
```

The code begins by importing csv, time, and numpy libraries. Function read_dict() converts a CSV (.csv) file to a list of OrderedDict elements. Function conv_reg_dict() converts a list of OrderedDict elements to a list of regular dictionary elements (for easier processing). Function sim_times() runs a simulation that creates two lists—lsd and lsgc. List lsd contains n run times for list comprension and list lsgc contains n run times for generator comprehension. Using simulation provides a more accurate picture of the true time it takes for both of these processes by running them over and over (n times). In this case, the simulation is run 1,000 times (n =1000). Of course, you can run the simulations as many or few times as you wish. Functions gen() and sim_gen() work together. Function gen() creates a generator. Function sim_gen() simulates the generator n times. I had to create these two functions because yielding a generator requires a different process than creating a generator comprehension. Function avg_ls() returns the mean (average) of a list of numbers. The main block begins by reading a CSV file (the one we created earlier in the chapter) into a list of OrderedDict elements, and converting it to a list of regular dictionary elements. The code continues by simulating run times of list comprehension and generator comprehension 1,000 times (n = 1000). The 1st simulation calculates 1,000 runtimes for traversing the dictionary list created earlier for both list and generator comprehension, and returns a list of those runtimes for each. The 2nd simulation calculates 1,000 runtimes by traversing the dictionary list for a generator, and returns a list of those runtimes. The code concludes by calculating the average runtime for each of the three techniques—list comprehension, generator comprehension, and generators—and comparing those averages.

The simulations verify that generator comprehension is more than ten times, and generators are more than eight times faster than list comprehension (runtimes will vary based on your PC). This makes sense because list comprehension stores all data in memory, while generators evaluate (lazily) as data is needed. Naturally, the speed advantage of generators becomes more important with big data sets. Without simulation, runtimes **cannot** be verified because we are randomly getting internal system clock times.

Data Randomization

A stochastic process is a family of random variables from some probability space into a state space (whew!). Simply, it is a random process through time. Data randomization is the process of selecting values from a sample in an unpredictable manner with the goal of simulating reality. Simulation allows application of data randomization in data science. The previous section demonstrated how simulation can be used to realistically compare iterables (list comprehension, generator comprehension, and generators).

In Python, pseudorandom numbers are used to simulate data randomness (reality). They are not truly random because the 1st generation has no previous number. We have to provide a seed (or random seed) to initialize a pseudorandom number generator. The random library implements pseudorandom number generators for various data distributions, and random.seed() is used to generate the initial (1st generation) seed number.

The following code example reads a CSV file and converts it to a list of regular dictionary elements. The code continues by creating a random number used to retrieve a random element from the list. Next, a generator of three randomly selected elements is created and displayed. The code continues by displaying three randomly shuffled elements from the list. The next section of code deterministically seeds the random number generator, which means that all generated random numbers will be the same based on the seed. So, the elements displayed will always be the same ones unless the seed is changed. The code then uses the system's time to nondeterministically generate random numbers and display those three elements. Next, nondeterministic random numbers are generated by another method and those three elements are displayed. The final part creates a names list so random choice and sampling methods can be used to display elements.

```
import csv, random, time

def read_dict(f, h):
    input_file = csv.DictReader(open(f), fieldnames=h)
    return (input_file)

def conv_reg_dict(d):
    return [dict(x) for x in d]

def r_inds(ls, n):
    length = len(ls) - 1
    yield [random.randrange(length) for _ in range(n)]

def get_slice(ls, n):
    return ls[:n]

def p_line():
    print ()
```

```python
if __name__ == '__main__':
    f = 'data/names.csv'
    headers = ['first', 'last']
    r_dict = read_dict(f, headers)
    dict_ls = conv_reg_dict(r_dict)
    n = len(dict_ls)
    r = random.randrange(0, n-1)
    print ('randomly selected index:', r)
    print ('randomly selected element:', dict_ls[r])
    elements = 3
    generator = next(r_inds(dict_ls, elements))
    p_line()
    print (elements, 'randomly generated indicies:', generator)
    print (elements, 'elements based on indicies:')
    for row in generator:
        print (dict_ls[row])
    x = [[i] for i in range(n-1)]
    random.shuffle(x)
    p_line()
    print ('1st', elements, 'shuffled elements:')
    ind = get_slice(x, elements)
    for row in ind:
        print (dict_ls[row[0]])
    seed = 1
    random_seed = random.seed(seed)
    rs1 = random.randrange(0, n-1)
    p_line()
    print ('deterministic seed', str(seed) + ':', rs1)
    print ('corresponding element:', dict_ls[rs1])
    t = time.time()
    random_seed = random.seed(t)
```

```python
rs2 = random.randrange(0, n-1)
p_line()
print ('non-deterministic time seed', str(t) + ' index:', rs2)
print ('corresponding element:', dict_ls[rs2], '\n')
print (elements, 'random elements seeded with time:')
for i in range(elements):
    r = random.randint(0, n-1)
    print (dict_ls[r], r)
random_seed = random.seed()
rs3 = random.randrange(0, n-1)
p_line()
print ('non-deterministic auto seed:', rs3)
print ('corresponding element:', dict_ls[rs3], '\n')
print (elements, 'random elements auto seed:')
for i in range(elements):
    r = random.randint(0, n-1)
    print (dict_ls[r], r)
names = []
for row in dict_ls:
    name = row['last'] + ', ' + row['first']
    names.append(name)
p_line()
print (elements, 'names with "random.choice()":')
for row in range(elements):
    print (random.choice(names))
p_line()
print (elements, 'names with "random.sample()":')
print (random.sample(names, elements))
```

Output:

```
randomly selected index: 85
randomly selected element: {'first': 'Heidi', 'last': 'Clare'}

3 randomly generated indicies: [10, 77, 136]
3 elements based on indicies:
{'first': 'Amanda', 'last': 'Lynn'}
{'first': 'Eaton', 'last': 'Wright'}
{'first': 'Rich', 'last': 'Mann'}

1st 3 shuffled elements:
{'first': 'Gene', 'last': 'Poole'}
{'first': 'Marty', 'last': 'Graw'}
{'first': 'Wanda', 'last': 'Rinn'}

deterministic seed 1: 34
corresponding element: {'first': 'April', 'last': 'Schauer'}

non-deterministic time seed 1512777603.6807067 index: 18
corresponding element: {'first': 'Anita', 'last': 'Job'}

3 random elements seeded with time:
{'first': 'Jay', 'last': 'Walker'} 96
{'first': 'Dick', 'last': 'Tator'} 70
{'first': 'Anita', 'last': 'Schhauer'} 23

non-deterministic auto seed: 127
corresponding element: {'first': 'Olive', 'last': 'Hoyl'}

3 random elements auto seed:
{'first': 'Royal', 'last': 'Payne'} 142
{'first': 'Harry', 'last': 'Legg'} 84
{'first': 'Ty', 'last': 'Knotts'} 161

3 names with "random.choice()":
Beard, Harry
Carr, Dusty
Gaiter, Ali

3 names with "random.sample()":
['Friese, Andy', 'Cade, Barry', 'Walker, Jay']
```

The code begins by importing csv, random, and time libraries. Functions read_dict() and conv_reg_dict() have already been explained. Function r_inds() generates a random list of n elements from the dictionary list. To get the proper length, one is subtracted because Python lists begin at index zero. Function get_slice() creates a randomly shuffled list of n elements from the dictionary list. Function p_line() prints a blank line. The main block begins by reading a CSV file and converting it into a list of regular dictionary elements. The code continues by creating a random number with random.randrange() based on the number of indices from the dictionary list, and displays the index and associated dictionary element. Next, a generator is created and populated with three randomly determined elements. The indices and associated elements are printed from the generator. The next part of the code randomly shuffles the indicies and puts them in list x. An index value is created by slicing three random elements based on the shuffled indices stored in list x. The three elements are then displayed. The code continues by creating a deterministic random seed using a fixed number (seed) in the function. So, the random number generated by this seed will be the same each time the program is run. This means that the dictionary element displayed will be also be the same. Next, two methods for creating nondeterministic random numbers are presented—random.seed(t) and random.seed()— where t varies by system time and using no parameter automatically varies random numbers. Randomly generated elements are displayed for each method. The final part of the code creates a list of names to hold just first and last names, so random.choice() and random.sample() can be used.

MongoDB and JSON

MongoDB is a document-based database classified as NoSQL. NoSQL (Not Only SQL database) is an approach to database design that can accommodate a wide variety of data models, including key-value, document, columnar, and graph formats. It uses JSON-like documents with schemas. It integrates extremely well with Python. A MongoDB collection is conceptually like a table in a relational database, and

27

a document is conceptually like a row. JSON is a lightweight data-interchange format that is easy for humans to read and write. It is also easy for machines to parse and generate.

Database queries from MongoDB are handled by PyMongo. PyMongo is a Python distribution containing tools for working with MongoDB. It is the most efficient tool for working with MongoDB using the utilities of Python. PyMongo was created to leverage the advantages of Python as a programming language and MongoDB as a database. The pymongo library is a native driver for MongoDB, which means it is it is it is built into Python language. Since it is native, the pymongo library is automatically available (doesn't have to be imported into the code).

The following code example reads a CSV file and converts it to a list of regular dictionary elements. The code continues by creating a JSON file from the dictionary list and saving it to disk. Next, the code connects to MongoDB and inserts the JSON data. The final part of the code manipulates data from the MongoDB database. First, all data in the database is queried and a few records are displayed. Second, the database is rewound. Rewind sets the pointer to back to the 1st database record. Finally, various queries are performed.

```python
import json, csv, sys, os
sys.path.append(os.getcwd()+'/classes')
import conn

def read_dict(f, h):
    input_file = csv.DictReader(open(f), fieldnames=h)
    return (input_file)

def conv_reg_dict(d):
    return [dict(x) for x in d]
```

```python
def dump_json(f, d):
    with open(f, 'w') as f:
        json.dump(d, f)

def read_json(f):
    with open(f) as f:
        return json.load(f)

if __name__ == '__main__':
    f = 'data/names.csv'
    headers = ['first', 'last']
    r_dict = read_dict(f, headers)
    dict_ls = conv_reg_dict(r_dict)
    json_file = 'data/names.json'
    dump_json(json_file, dict_ls)
    data = read_json(json_file)
    obj = conn.conn('test')
    db = obj.getDB()
    names = db.names
    names.drop()
    for i, row in enumerate(data):
        row['_id'] = i
        names.insert_one(row)
    n = 3
    print('1st', n, 'names:')
    people = names.find()
    for i, row in enumerate(people):
        if i < n:
            print (row)
    people.rewind()
    print('\n1st', n, 'names with rewind:')
    for i, row in enumerate(people):
```

```
    if i < n:
        print (row)
print ('\nquery 1st', n, 'names')
first_n = names.find().limit(n)
for row in first_n:
    print (row)
print ('\nquery last', n, 'names')
length = names.find().count()
last_n = names.find().skip(length - n)
for row in last_n:
    print (row)
fnames = ['Ella', 'Lou']
lnames = ['Vader', 'Pole']
print ('\nquery Ella:')
query_1st_in_list = names.find( {'first':{'$in':[fnames[0]]}})
for row in query_1st_in_list:
    print (row)
print ('\nquery Ella or Lou:')
query_1st = names.find( {'first':{'$in':fnames}} )
for row in query_1st:
    print (row)
print ('\nquery Lou Pole:')
query_and = names.find( {'first':fnames[1], 'last':lnames[1]} )
for row in query_and:
    print (row)
print ('\nquery first name Ella or last name Pole:')
query_or = names.find( {'$or':[{'first':fnames[0]},
{'last':lnames[1]}]} )
```

```
for row in query_or:
    print (row)
pattern = '^Sch'
print ('\nquery regex pattern:')
query_like = names.find( {'last':{'$regex':pattern}} )
for row in query_like:
    print (row)
pid = names.count()
doc = {'_id':pid, 'first':'Wendy', 'last':'Day'}
names.insert_one(doc)
print ('\ndisplay added document:')
q_added = names.find({'first':'Wendy'})
print (q_added.next())
print ('\nquery last n documents:')
q_n = names.find().skip((pid-n)+1)
for _ in range(n):
    print (q_n.next())
```

Class conn:

```
class conn:
    from pymongo import MongoClient
    client = MongoClient('localhost', port=27017)
    def __init__(self, dbname):
        self.db = conn.client[dbname]
    def getDB(self):
        return self.db
```

Output:

```
1st 3 names:
{'_id': 0, 'first': 'Adam', 'last': 'Baum'}
{'_id': 1, 'first': 'Adam', 'last': 'Zapel'}
{'_id': 2, 'first': 'Al', 'last': 'Bino'}

1st 3 names with rewind:
{'_id': 0, 'first': 'Adam', 'last': 'Baum'}
{'_id': 1, 'first': 'Adam', 'last': 'Zapel'}
{'_id': 2, 'first': 'Al', 'last': 'Bino'}

query 1st 3 names
{'_id': 0, 'first': 'Adam', 'last': 'Baum'}
{'_id': 1, 'first': 'Adam', 'last': 'Zapel'}
{'_id': 2, 'first': 'Al', 'last': 'Bino'}

query last 3 names
{'_id': 163, 'first': 'Will', 'last': 'Power'}
{'_id': 164, 'first': 'Willie', 'last': 'Waite'}
{'_id': 165, 'first': 'Willie', 'last': 'Makeit'}

query Ella:
{'_id': 79, 'first': 'Ella', 'last': 'Vader'}

query Ella or Lou:
{'_id': 79, 'first': 'Ella', 'last': 'Vader'}
{'_id': 108, 'first': 'Lou', 'last': 'Pole'}

query Lou Pole:
{'_id': 108, 'first': 'Lou', 'last': 'Pole'}

query first name Ella or last name Pole:
{'_id': 79, 'first': 'Ella', 'last': 'Vader'}
{'_id': 108, 'first': 'Lou', 'last': 'Pole'}

query regex pattern:
{'_id': 23, 'first': 'Anita', 'last': 'Schhauer'}
{'_id': 34, 'first': 'April', 'last': 'Schauer'}

display added document:
{'_id': 166, 'first': 'Wendy', 'last': 'Day'}

query last n documents:
{'_id': 164, 'first': 'Willie', 'last': 'Waite'}
{'_id': 165, 'first': 'Willie', 'last': 'Makeit'}
{'_id': 166, 'first': 'Wendy', 'last': 'Day'}
```

The code begins by importing json, csv, sys, and os libraries. Next, a path (sys.path.append) to the class conn is established. Method getcwd() (from the os library) gets the current working directory for classes. Class conn is then imported. I built this class to simplify connectivity to the database from any program. The code continues with four functions. Functions read_dict() and conv_reg_dict() were explained earlier. Function dump_json() writes JSON data to disk. Function read_json() reads JSON data from disk. The main block begins by reading a CSV file and converting it into a list of regular dictionary elements. Next, the list is dumped to disk as JSON. The code continues by creating a PyMongo connection instance test as an object and assigning it to variable obj. You can create any instance you wish, but test is the default. Next, the database instance is assigned to db by method getDB() from obj. Collection names is then created in MongoDB and assigned to variable names. When prototyping, I always drop the collection before manipulating it. This eliminates duplicate key errors. The code continues by inserting the JSON data into the collection. For each document in a MongoDB collection, I explicitly create primary key values by assigning sequential numbers to _id. MongoDB exclusively uses _id as the primary key identifier for each document in a collection. If you don't name it yourself, a system identifier is automatically created, which is messy to work with in my opinion. The code continues with PyMongo query names.find(), which retrieves all documents from the names collection. Three records are displayed just to verify that the query is working. To reuse a query that has already been accessed, rewind() must be issued. The next PyMongo query accesses and displays three (n = 3) documents. The next query accesses and displays the last three documents. Next, we move into more complex queries. First, access documents with first name Ella. Second, access documents with first names Ella or Lou. Third, access document Lou Pole. Fourth, access documents with first name Ella or last name Pole. Next, a regular expression is used to access documents with last names beginning with

Sch. A regular expression is a sequence of characters that define a search pattern. Finally, add a new document, display it, and display the last three documents in the collection.

Visualization

Visualization is the process of representing data graphically and leveraging these representations to gain insight into the data. Visualization is one of the most important skills in data science because it facilitates the way we process large amounts of complex data.

The following code example creates and plots a normally distributed set of data. It then shifts data to the left (and plots) and shifts data to the right (and plots). A normal distribution is a probability distribution that is symmetrical about the mean, and is very important to data science because it is an excellent model of how events naturally occur in reality.

```
import matplotlib.pyplot as plt
from scipy.stats import norm
import numpy as np

if __name__ == '__main__':
    x = np.linspace(norm.ppf(0.01), norm.ppf(0.99), num=100)
    x_left = x - 1
    x_right = x + 1
    y = norm.pdf(x)
    plt.ylim(0.02, 0.41)
    plt.scatter(x, y, color='crimson')
    plt.fill_between(x, y, color='crimson')
    plt.scatter(x_left, y, color='chartreuse')
    plt.scatter(x_right, y, color='cyan')
    plt.show()
```

Output:

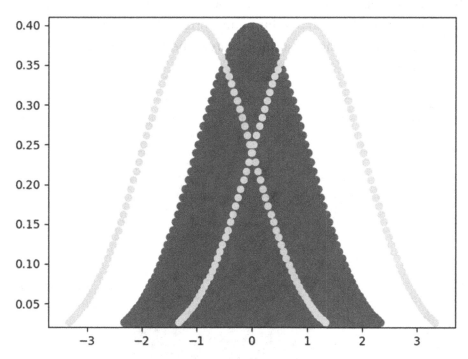

Figure 1-1. *Normally distributed data*

The code example (Figure 1-1) begins by importing matplotlib, scipy, and numpy libraries. The matplotlib library is a 2-D plotting module that produces publication quality figures in a variety of hardcopy formats and interactive environments across platforms. The SciPy library provides user-friendly and efficient numerical routines for numerical integration and optimization. The main block begins by creating a sequence of 100 numbers between 0.01 and 0.99. The reason is the normal distribution is based on probabilities, which must be between zero and one. The code continues by shifting the sequence one unit to the left and one to the right for later plotting. The ylim() method is used to pull the chart to the bottom (x-axis). A scatter plot is created for the original data, one unit to the left, and one to the right, with different colors for effect.

On the 1st line of the main block in the linespace() function, increase the number of data points from num = 100 to num = 1000 and see what happens. The result is a smoothing of the normal distribution, because more data provides a more realistic picture of the natural world.

Output:

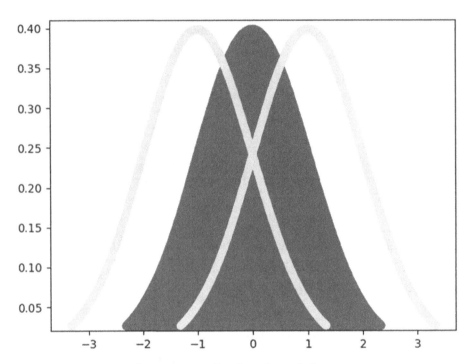

Figure 1-2. *Smoothing normally distributed data*

Smoothing works (Figure 1-2) because a normal distribution consists of continuous random variables. A continuous random variable is a random variable with a set of infinite and uncountable values. So, more data creates more predictive realism. Since we cannot add infinite data, we work with as much data as we can. The tradeoff is more data increases computer processing resources and execution time. Data scientists must thereby weigh this tradeoff when conducting their tradecraft.

CHAPTER 2

Monte Carlo Simulation and Density Functions

Monte Carlo simulation (MCS) applies repeated random sampling (randomness) to obtain numerical results for deterministic problem solving. It is widely used in optimization, numerical integration, and risk-based decision making. Probability and cumulative density functions are statistical measures that apply probability distributions for random variables, and can be used in conjunction with MCS to solve deterministic problem.

Note Reader can refer to the download source code file to see color figs in this chapter.

Stock Simulations

The 1st example is hypothetical and simple, but useful in demonstrating data randomization. It begins with a fictitious stock priced at $20. It then projects price out 200 days and plots.

© David Paper 2018
D. Paper, *Data Science Fundamentals for Python and MongoDB*,
https://doi.org/10.1007/978-1-4842-3597-3_2

```python
import matplotlib.pyplot as plt, numpy as np
from scipy import stats

def cum_price(p, d, m, s):
    data = []
    for d in range(d):
            prob = stats.norm.rvs(loc=m, scale=s)
            price = (p * prob)
            data.append(price)
            p = price
    return data

if __name__ == "__main__":
    stk_price, days, mean, s = 20, 200, 1.001, 0.005
    data = cum_price(stk_price, days, mean, s)
    plt.plot(data, color='lime')
    plt.ylabel('Price')
    plt.xlabel('days')
    plt.title('stock closing prices')
    plt.show()
```

Output:

Figure 2-1. *Simple random plot*

The code begins by importing matplotlib, numpy, and scipy libraries. It continues with function cum_price(), which generates 200 normally distributed random numbers (one for each day) with norm_rvs(). Data randomness is key. The main block creates the variables. Mean is set a bit over 1 and standard deviation (s) at a very small number to generate a slowly increasing stock price. Mean (mu) is the average change in value. Standard deviation is the variation or dispersion in the data. With s of 0.005, our data has very little variation. That is, the numbers in our data set are very close to each other. Remember that this is not a real scenario! The code continues by plotting results as shown in Figure 2-1.

The next example adds MCS into the mix with a while loop that iterates 100 times:

```
import matplotlib.pyplot as plt, numpy as np
from scipy import stats

def cum_price(p, d, m, s):
    data = []
    for d in range(d):
            prob = stats.norm.rvs(loc=m, scale=s)
            price = (p * prob)
            data.append(price)
            p = price
    return data

if __name__ == "__main__":
    stk_price, days, mu, sigma = 20, 200, 1.001, 0.005
    x = 0
    while x < 100:
        data = cum_price(stk_price, days, mu, sigma)
        plt.plot(data)
        x += 1
    plt.ylabel('Price')
    plt.xlabel('day')
    plt.title('Stock closing price')
    plt.show()
```

Output:

Figure 2-2. *Monte Carlo simulation augmented plot*

The while loop allows us to visualize (as shown in Figure 2-2) 100 possible stock price outcomes over 200 days. Notice that mu (mean) and sigma (standard deviation) are used. This example demonstrates the power of MCS for decision making.

What-If Analysis

What-If analysis changes values in an algorithm to see how they impact outcomes. Be sure to only change one variable at a time, otherwise you won't know which caused the change. In the previous example, what if we change days to 500 while keeping all else constant (the same)? Plotting this change results in the following (Figure 2-3):

Figure 2-3. *What-If analysis for 500 days*

Notice that the change in price is slower. Changing mu (mean) to 1.002 (don't forget to change days back to 200) results in faster change (larger averages) as follows (Figure 2-4):

Figure 2-4. *What-If analysis for mu = 1.002*

Changing sigma to 0.02 results in more variation as follows (Figure 2-5):

Figure 2-5. *What-If analysis for sigma = 0.02*

Product Demand Simulation

A discrete probability is the probability of each discrete random value
occurring in a sample space or population. A random variable assumes
different values determined by chance. A discrete random variable can
only assume a countable number of values. In contrast, a continuous
random variable can assume an uncountable number of values in a line
interval such as a normal distribution.

In the code example, demand for a fictitious product is predicted by
four discrete probability outcomes: 10% that random variable is 10,000
units, 35% that random variable is 20,000 units, 30% that random variable
is 40,000 units, and 25% that random variable is 60,000 units. Simply,

10% of the time demand is 10,000, 35% of the time demand is 20,000, 30% of the time demand is 40,000, and 25% of the time demand is 60,000. Discrete outcomes must total 100%. The code runs MCS on a production algorithm that determines profit for each discrete outcome, and plots the results.

```python
import matplotlib.pyplot as plt, numpy as np

def demand():
    p = np.random.uniform(0,1)
    if p < 0.10:
        return 10000
    elif p >= 0.10 and p < 0.45:
        return 20000
    elif p >= 0.45 and p < 0.75:
        return 40000
    else:
        return 60000

def production(demand, units, price, unit_cost, disposal):
    units_sold = min(units, demand)
    revenue = units_sold * price
    total_cost = units * unit_cost
    units_not_sold = units - demand
    if units_not_sold > 0:
        disposal_cost = disposal * units_not_sold
    else:
        disposal_cost = 0
    profit = revenue - total_cost - disposal_cost
    return profit
```

```python
def mcs(x, n, units, price, unit_cost, disposal):
    profit = []
    while x <= n:
        d = demand()
        v = production(d, units, price, unit_cost, disposal)
        profit.append(v)
        x += 1
    return profit

def max_bar(ls):
    tup = max(enumerate(ls))
    return tup[0] - 1

if __name__ == "__main__":
    units = [10000, 20000, 40000, 60000]
    price, unit_cost, disposal = 4, 1.5, 0.2
    avg_p = []
    x, n = 1, 10000
    profit_10 = mcs(x, n, units[0], price, unit_cost, disposal)
    avg_p.append(np.mean(profit_10))
    print ('Profit for {:,.0f}'.format(units[0]),
           'units: ${:,.2f}'.format(np.mean(profit_10)))
    profit_20 = mcs(x, n, units[1], price, unit_cost, disposal)
    avg_p.append(np.mean(np.mean(profit_20)))
    print ('Profit for {:,.0f}'.format(units[1]),
           'units: ${:,.2f}'.format(np.mean(profit_20)))
    profit_40 = mcs(x, n, units[2], price, unit_cost, disposal)
    avg_p.append(np.mean(profit_40))
    print ('Profit for {:,.0f}'.format(units[2]),
           'units: ${:,.2f}'.format(np.mean(profit_40)))
    profit_60 = mcs(x, n, units[3], price, unit_cost, disposal)
    avg_p.append(np.mean(profit_60))
```

```
print ('Profit for {:,.0f}'.format(units[3]),
      'units: ${:,.2f}'.format(np.mean(profit_60)))
labels = ['10000','20000','40000','60000']
pos = np.arange(len(labels))
width = 0.75 # set less than 1.0 for spaces between bins
plt.figure(2)
ax = plt.axes()
ax.set_xticks(pos + (width / 2))
ax.set_xticklabels(labels)
barlist = plt.bar(pos, avg_p, width, color='aquamarine')
barlist[max_bar(avg_p)].set_color('orchid')
plt.ylabel('Profit')
plt.xlabel('Production Quantity')
plt.title('Production Quantity by Demand')
plt.show()
```

Output:

```
Profit for 10,000 units: $25,000.00
Profit for 20,000 units: $45,829.40
Profit for 40,000 units: $57,886.60
Profit for 60,000 units: $44,882.40
```

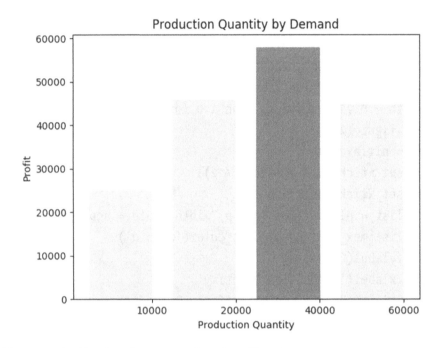

Figure 2-6. *Production quantity visualization*

The code begins by importing matplotlib and numpy libraries. It continues with four functions. Function demand() begins by randomly generating a uniformly distributed probability. It continues by returning one of the four discrete probability outcomes established by the problem we wish to solve. Function production() returns profit based on an algorithm that I devised. Keep in mind that any profit-base algorithm can be substitued, which illuminates the incredible flexibility of MCS. Function mcs() runs the simulation 10,000 times. Increasing the number of runs provides better prediction accuracy with costs being more computer processing resources and runtime. Function max_bar() establishes the highest bar in the bar chart for better illumination. The main block begins by simulating profit for each discrete probability outcome, and printing and visualizing results. MCS predicts that production quantity of 40,000 units yields the highest profit, as shown in Figure 2-6.

Increasing the number of MCS simulations results in a more accurate prediction of reality because it is based on stochastic reasoning (data randomization). You can also substitute any discrete probability distribution based on your problem-solving needs with this code structure. As alluded to earlier, you can use any algorithm you wish to predict with MCS, making it an incredibly flexible tool for data scientists.

We can further enhance accuracy by running an MCS on an MCS. The code example uses the same algorithm and process as before, but adds an MCS on the original MCS to get a more accurate prediction:

```python
import matplotlib.pyplot as plt, numpy as np

def demand():
    p = np.random.uniform(0,1)
    if p < 0.10:
        return 10000
    elif p >= 0.10 and p < 0.45:
        return 20000
    elif p >= 0.45 and p < 0.75:
        return 40000
    else:
        return 60000

def production(demand, units, price, unit_cost, disposal):
    units_sold = min(units, demand)
    revenue = units_sold * price
    total_cost = units * unit_cost
    units_not_sold = units - demand
    if units_not_sold > 0:
        disposal_cost = disposal * units_not_sold
```

```python
    else:
        disposal_cost = 0
    profit = revenue - total_cost - disposal_cost
    return profit

def mcs(x, n, units, price, unit_cost, disposal):
    profit = []
    while x <= n:
        d = demand()
        v = production(d, units, price, unit_cost, disposal)
        profit.append(v)
        x += 1
    return profit

def display(p, i):
    print ('Profit for {:,.0f}'.format(units[i]),
            'units: ${:,.2f}'.format(np.mean(p)))

if __name__ == "__main__":
    units = [10000, 20000, 40000, 60000]
    price, unit_cost, disposal = 4, 1.5, 0.2
    avg_ls = []
    x, n, y, z = 1, 10000, 1, 1000
    while y <= z:
        profit_10 = mcs(x, n, units[0], price, unit_cost,
        disposal)
        profit_20 = mcs(x, n, units[1], price, unit_cost,
        disposal)
        avg_profit = np.mean(profit_20)
        profit_40 = mcs(x, n, units[2], price, unit_cost,
        disposal)
        avg_profit = np.mean(profit_40)
        profit_60 = mcs(x, n, units[3], price, unit_cost,
        disposal)
```

```
avg_profit = np.mean(profit_60)
avg_ls.append({'p10':np.mean(profit_10),
               'p20':np.mean(profit_20),
               'p40':np.mean(profit_40),
               'p60':np.mean(profit_60)})
    y += 1
mcs_p10, mcs_p20, mcs_p40, mcs_p60 = [], [], [], []
for row in avg_ls:
    mcs_p10.append(row['p10'])
    mcs_p20.append(row['p20'])
    mcs_p40.append(row['p40'])
    mcs_p60.append(row['p60'])
display(np.mean(mcs_p10), 0)
display(np.mean(mcs_p20), 1)
display(np.mean(mcs_p40), 2)
display(np.mean(mcs_p60), 3)
```

Output:

```
Profit for 10,000 units: $25,000.00
Profit for 20,000 units: $45,800.24
Profit for 40,000 units: $57,980.97
Profit for 60,000 units: $44,996.88
```

The code for this example is the same as the previous one, except for the MCS while loop (while $y <= z$). In this loop, profits are calculated as before using function mcs(), but each simulation result is appended to list avg_ls. So, avg_ls contains 1,000 ($z = 1000$) simulation results of the original simulation results. Accuracy is increased, but more computer resources and runtime are required. Running 1,000 simulations on the original MCS takes a bit over one minute, which is a lot of processing time!

Randomness Using Probability and Cumulative Density Functions

Randomness masquerades as reality (the natural world) in data science, since the future cannot be predicted. That is, randomization is the way data scientists simulate reality. More data means better accuracy and prediction (more realism). It plays a key role in discrete event simulation and deterministic problem solving. Randomization is used in many fields such as statistics, MCS, cryptography, statistics, medicine, and science.

The density of a continuous random variable is its probability density function (PDF). PDF is the probability that a random variable has the value x, where x is a point within the interval of a sample. This probability is determined by the integral of the random variable's PDF over the range (interval) of the sample. That is, the probability is given by the area under the density function, but above the horizontal axis and between the lowest and highest values of range. An integral (integration) is a mathematical object that can be interpreted as an area under a normal distribution curve. A cumulative distribution function (CDF) is the probability that a random variable has a value less than or equal to x. That is, CDF accumulates all of the probabilities less than or equal to x. The percent point function (PPF) is the inverse of the CDF. It is commonly referred to as the inverse cumulative distribution function (ICDF). ICDF is very useful in data science because it is the actual value associated with an area under the PDF. Please refer to www.itl.nist.gov/div898/handbook/eda/section3/eda362.htm for an excellent explanation of density functions.

As stated earlier, a probability is determined by the integral of the random variable's PDF over the interval of a sample. That is, integrals are used to determine the probability of some random variable falling within a certain range (sample). In calculus, the integral represents a class of functions (the antiderivative) whose derivative is the integrand. The integral symbol represents integration, while an integrand is the function being integrated in either a definite or indefinite integral. The fundamental theorem of calculus relates the evaluation of definitive integrals to indefinite integrals. The only reason I include this information here is to emphasize the importance of calculus to data science. Another aspect of calculus important to data science, "gradient descent," is presented later in Chapter 4.

Although theoretical explanations are invaluable, they may not be intuitive. A great way to better understand these concepts is to look at an example.

In the code example, 2-D charts are created for PDF, CDF, and ICDF (PPF). The idea of a colormap is included in the example. A colormap is a lookup table specifying the colors to be used in rendering palettized image. A palettized image is one that is efficiently encoded by mapping its pixels to a palette containing only those colors that are actually present in the image. The matplotlib library includes a myriad of colormaps. Please refer to https://matplotlib.org/examples/color/colormaps_reference.html for available colormaps.

```
import matplotlib.pyplot as plt
from scipy.stats import norm
import numpy as np

if __name__ == '__main__':
    x = np.linspace(norm.ppf(0.01), norm.ppf(0.99), num=1000)
    y1 = norm.pdf(x)
    plt.figure('PDF')
    plt.xlim(x.min()-.1, x.max()+0.1)
    plt.ylim(y1.min(), y1.max()+0.01)
    plt.xlabel('x')
    plt.ylabel('Probability Density')
    plt.title('Normal PDF')
    plt.scatter(x, y1, c=x, cmap='jet')
```

```
plt.fill_between(x, y1, color='thistle')
plt.show()
plt.close('PDF')
plt.figure('CDF')
plt.xlabel('x')
plt.ylabel('Probability')
plt.title('Normal CDF')
y2 = norm.cdf(x)
plt.scatter(x, y2, c=x, cmap='jet')
plt.show()
plt.close('CDF')
plt.figure('ICDF')
plt.xlabel('Probability')
plt.ylabel('x')
plt.title('Normal ICDF (PPF)')
y3 = norm.ppf(x)
plt.scatter(x, y3, c=x, cmap='jet')
plt.show()
plt.close('ICDF')
```

Output:

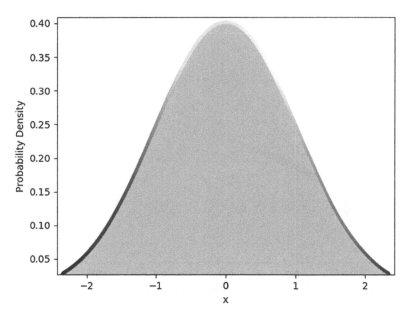

Figure 2-7. *Normal probability density function visualization*

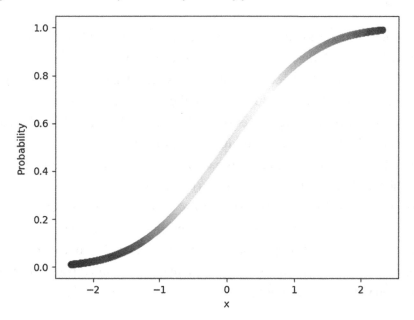

Figure 2-8. *Normal cumulative distribution function visualization*

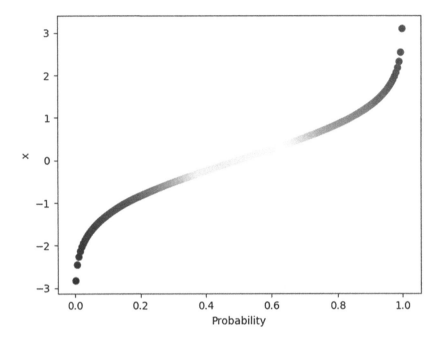

Figure 2-9. *Normal inverse cumulative distribution function*
visualization

The code begins by importing three libraries–matplotlib, scipy, and
numpy. The main block begins by creating a sequence of 1,000 x values
between 0.01 and 0.99 (because probabilities must fall between 0 and 1).
Next, a sequence of PDF y values is created based on the x values. The
code continues by plotting the resultant PDF shown in Figure 2-7. Next, a
sequence of CDF (Figure 2-8) and ICDF (Figure 2-9) values are created and
plotted. From the visualization, it is easier to see that the PDF represents
all of the possible x values (probabilities) that exist under the normal
distribution. It is also easier to visualize the CDF because it represents
the accumulation of all the possible probabilities. Finally, the ICDF is
easier to understand through visualization (see Figure 2-9) because the
x-axis represents probabilities, while the y-axis represents the actual value
associated with those probabilities.

Let's apply ICDF. Suppose you are a data scientist at Apple and your boss asks you to determine Apple iPhone 8 failure rates so she can develop a mockup presentation for her superiors. For this hypothetical example, your boss expects four calculations: time it takes 5% of phones to fail, time interval (range) where 95% of phones fail, time where 5% of phones survive (don't fail), and time interval where 95% of phones survive. In all cases, report time in hours. From data exploration, you ascertain average (mu) failure time is 1,000 hours and standard deviation (sigma) is 300 hours.

The code example calculates ICDF for the four scenarios and displays the results in an easy to understand format for your boss:

```
from scipy.stats import norm
import numpy as np

def np_rstrip(v):
    return np.char.rstrip(v.astype(str), '.0')

def transform(t):
    one, two = round(t[0]), round(t[1])
    return (np_rstrip(one), np_rstrip(two))

if __name__ == "__main__":
    mu, sigma = 1000, 300
    print ('Expected failure rates:')
    fail = np_rstrip(round(norm.ppf(0.05, loc=mu, scale=sigma)))
    print ('5% fail within', fail, 'hours')
    fail_range = norm.interval(0.95, loc=mu, scale=sigma)
    lo, hi = transform(fail_range)
    print ('95% fail between', lo, 'and', hi, end=' ')
    print ('hours of usage')
    print ('\nExpected survival rates:')
    last = np_rstrip(round(norm.ppf(0.95, loc=mu, scale=sigma)))
    print ('5% survive up to', last, 'hours of usage')
```

```
last_range = norm.interval(0.05, loc=mu, scale=sigma)
lo, hi = transform(last_range)
print ('95% survive between', lo, 'and', hi, 'hours of usage')
```

Output:

```
Expected failure rates:
5% fail within 507 hours
95% fail between 412 and 1588 hours of usage

Expected survival rates:
5% survive up to 1493 hours of usage
95% survive between 981 and 1019 hours of usage
```

The code example begins by importing scipy and numpy libraries. It continues with two functions. Function np_rstrip() converts numpy float to string and removes extraneous characters. Function transform() rounds and returns a tuple. Both are just used to round numbers to no decimal places to make it user-friendly for your fictitious boss. The main block begins by initializing mu and sigma to 1,000 (failures) and 300 (variates). That is, on average, our smartphones fail within 1,000 hours, and failures vary between 700 and 1,300 hours. Next, find the ICDF value for a 5% failure rate and an interval where 95% fail with norm.ppf(). So, 5% of all phones are expected to fail within 507 hours, while 95% fail between 412 and 1,588 hours of usage. Next, find the ICDF value for a 5% survival rate and an interval where 95% survive. So, 5% of all phones survive up to 1,493 hours, while 95% survive between 981 and 1,019 hours of usage.

Simply, ICDF allows you to work backward from a known probability to find an x value! Please refer to http://support.minitab.com/en-us/minitab-express/1/help-and-how-to/basic-statistics/probability-distributions/supporting-topics/basics/using-the-inverse-cumulative-distribution-function-icdf/#what-is-an-inverse-cumulative-distribution-function-icdf for more information.

Let's try What-if analysis. What if we reduce error rate (sigma) from 300 to 30?

```
Expected failure rates:
5% fail within 951 hours
95% fail between 941 and 1059 hours of usage

Expected survival rates:
5% survive up to 1049 hours of usage
95% survive between 998 and 1002 hours of usage
```

Now, 5% of all phones are expected to fail within 951 hours, while 95% fail between 941 and 1,059 hours of usage. And, 5% of all phones survive up to 1,049 hours, while 95% survive between 998 and 1,002 hours of usage. What does this mean? Less variation (error) shows that values are much closer to the average for both failure and survival rates. This makes sense because variation is calculated from a mean of 1,000.

Let's shift to a simulation example. Suppose your boss asks you to find the optimal monthly order quantity for a type of car given that demand is normally distributed (it must, because PDF is based on this assumption), average demand (mu) is 200, and variation (sigma) is 30. Each car costs $25,000, sells for $45,000, and half of the cars not sold at full price can be sold for $30,000. Like other MCS experiments, you can modify the profit algorithm to enhance realism. By suppliers, you are limited to order quantities of 160, 180, 200, 220, 240, 260, or 280.

MCS is used to find the profit for each order based on the information provided. Demand is generated randomly for each iteration of the simulation. Profit calculations by order are automated by running MCS for each order.

```
import numpy as np
import matplotlib.pyplot as plt

def str_int(s):
    val = "%.2f" % profit
    return float(val)
```

```python
if __name__ == "__main__":
    orders = [180, 200, 220, 240, 260, 280, 300]
    mu, sigma, n = 200, 30, 10000
    cost, price, discount = 25000, 45000, 30000
    profit_ls = []
    for order in orders:
        x = 1
        profit_val = []
        inv_cost = order * cost
        while x <= n:
            demand = round(np.random.normal(mu, sigma))
            if demand < order:
                diff = order - demand
                if diff > 0:
                    damt = round(abs(diff) / 2) * discount
                    profit = (demand * price) - inv_cost + damt
                else:
                    profit = (order * price) - inv_cost
            else:
                profit = (order * price) - inv_cost
            profit = str_int(profit)
            profit_val.append(profit)
            x += 1
        avg_profit = np.mean(profit_val)
        profit_ls.append(avg_profit)
        print ('${0:,.2f}'.format(avg_profit), '(profit)',
                'for order:', order)
    max_profit = max(profit_ls)
    profit_np = np.array(profit_ls)
    max_ind = np.where(profit_np == profit_np.max())
    print ('\nMaximum profit', '${0:,.2f}'.format(max_profit),
```

```
        'for order', orders[int(max_ind[0])])
    barlist = plt.bar(orders, profit_ls, width=15,
color='thistle')
    barlist[int(max_ind[0])].set_color('lime')
    plt.title('Profits by Order Quantity')
    plt.xlabel('orders')
    plt.ylabel('profit')
    plt.tight_layout()
    plt.show()
```

Output:

```
$3,460,479.00 (profit) for order: 180
$3,638,933.50 (profit) for order: 200
$3,669,597.50 (profit) for order: 220
$3,554,889.00 (profit) for order: 240
$3,385,369.00 (profit) for order: 260
$3,200,210.00 (profit) for order: 280
$2,994,411.00 (profit) for order: 300

Maximum profit $3,669,597.50 for order 220
```

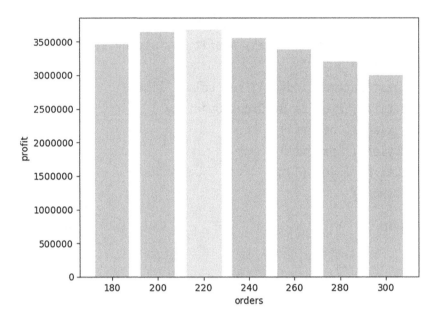

Figure 2-10. *Profits by order quantity visualization*

The code begins by importing numpy and matplotlib. It continues with a function (str_int()) that converts a string to float. The main block begins by initializing orders, mu, sigma, n, cost, price, discount, and list of profits by order. It continues by looping through each order quantity and running MCS with 10,000 iterations. A randomly generated demand probability is used to calculate profit for each iteration of the simulation. The technique for calculating profit is pretty simple, but you can substitute your own algorithm. You can also modify any of the given information based on your own data. After calculating profit for each order through MCS, the code continues by finding the order quantity with the highest profit. Finally, the code generates a bar chart to illuminate results though visualization shown in Figure 2-10.

The final code example creates a PDF visualization:

```python
import matplotlib.pyplot as plt, numpy as np
from scipy.stats import norm

if __name__ == '__main__':
    n = 100
    x = np.linspace(norm.ppf(0.01), norm.ppf(0.99), num=n)
    y = norm.pdf(x)
    dic = {}
    for i, row in enumerate(y):
        dic[x[i]] = [np.random.uniform(0, row) for _ in range(n)]
    xs = []
    ys = []
    for key, vals in dic.items():
        for y in vals:
            xs.append(key)
            ys.append(y)
    plt.xlim(min(xs), max(xs))
    plt.ylim(0, max(ys)+0.02)
    plt.title('Normal PDF')
    plt.xlabel('x')
    plt.ylabel('Probability Density')
    plt.scatter(xs, ys, c=xs, cmap='rainbow')
    plt.show()
```

Output:

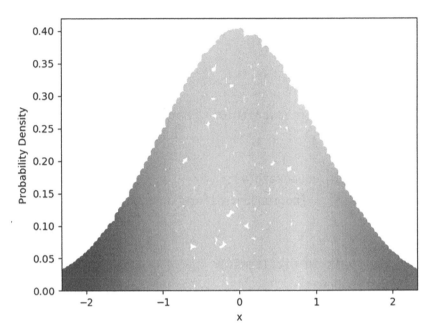

Figure 2-11. *All PDF probabilities with 100 simulations*

The code begins by importing matplotlib, numpy, and scipy libraries. The main block begins by initializing the number of points you wish to plot, PDF x and y values, and a dictionary. To plot all PDF probabilities, a set of randomly generated values for each point on the x-axis is created. To accomplish this task, the code assigns 100 (n = 100) values to x from 0.01 to 0.99. It continues by assigning 100 PDF values to y. Next, a dictionary element is populated by a (key, value) pair consisting of each x value as key and a list of 100 (n = 100) randomly generated numbers between 0 and pdf(x) as value associated with x. Although the code creating the dictionary is simple, please think carefully about what is happening because it is pretty abstract. The code continues by building (x, y) pairs from the dictionary. The result is 10,000 (100 X 100) (x, y) pairs, where each 100 x values has 100 associated y values visualized in Figure 2-11.

To smooth out the visualization increase n to 1,000 (n = 1000) at the beginning of the main block:

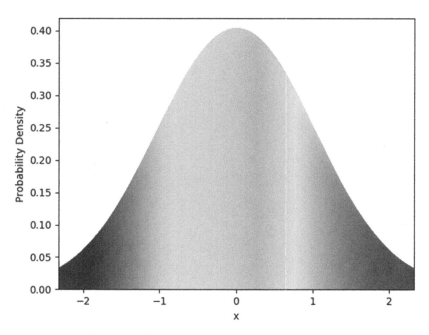

Figure 2-12. *All PDF probabilities with 1,000 simulations*

By increasing n to 1000, 1,000,000 (1,000 X 1,000) (x, y) pairs are plotted as shown in Figure 2-12!

CHAPTER 3

Linear Algebra

Linear algebra is a branch of mathematics concerning vector spaces and linear mappings between such spaces. Simply, it explores linelike relationships. Practically every area of modern science approximates modeling equations with linear algebra. In particular, data science relies on linear algebra for machine learning, mathematical modeling, and dimensional distribution problem solving.

Vector Spaces

A vector space is a collection of vectors. A vector is any quantity with magnitude and direction that determines the position of one point in space relative to another. Magnitude is the size of an object measured by movement, length, and/or velocity. Vectors can be added and multiplied (by scalars) to form new vectors. A scalar is any quantity with magnitude (size). In application, vectors are points in finite space.

Vector examples include breathing, walking, and displacement. Breathing requires diaphragm muscles to exert a force that has magnitude and direction. Walking requires movement in some direction. Displacement measures how far an object moves in a certain direction.

© David Paper 2018
D. Paper, *Data Science Fundamentals for Python and MongoDB*,
https://doi.org/10.1007/978-1-4842-3597-3_3

Vector Math

In vector math, a vector is depicted as a directed line segment whose length is its magnitude vector with an arrow indicating direction from tail to head. Tail is where the line segment begins and head is where it ends (the arrow). Vectors are the same if they have the same magnitude and direction.

To add two vectors a and b, start b where a finishes, and complete the triangle. Visually, start at some point of origin, draw a (Figure 3-1), start b (Figure 3-2) from head of a, and the result c (Figure 3-3) is a line from tail of a to head of b. The 1st example illustrates vector addition as well as a graphic depiction of the process:

```python
import matplotlib.pyplot as plt, numpy as np

def vector_add(a, b):
    return np.add(a, b)

def set_up():
    plt.figure()
    plt.xlim(-.05, add_vectors[0]+0.4)
    plt.ylim(-1.1, add_vectors[1]+0.4)

if __name__ == "__main__":
    v1, v2 = np.array([3, -1]), np.array([2, 3])
    add_vectors = vector_add(v1, v2)
    set_up()
    ax = plt.axes()
    ax.arrow(0, 0, 3, -1, head_width=0.1, fc='b', ec='b')
    ax.text(1.5, -0.35, 'a')
    ax.set_facecolor('honeydew')
    set_up()
    ax = plt.axes()
    ax.arrow(0, 0, 3, -1, head_width=0.1, fc='b', ec='b')
    ax.arrow(3, -1, 2, 3, head_width=0.1, fc='crimson',
    ec='crimson')
```

```
ax.text(1.5, -0.35, 'a')
ax.text(4, -0.1, 'b')
ax.set_facecolor('honeydew')
set_up()
ax = plt.axes()
ax.arrow(0, 0, 3, -1, head_width=0.1, fc='b', ec='b')
ax.arrow(3, -1, 2, 3, head_width=0.1, fc='crimson',
ec='crimson')
ax.arrow(0, 0, 5, 2, head_width=0.1, fc='springgreen',
ec='springgreen')
ax.text(1.5, -0.35, 'a')
ax.text(4, -0.1, 'b')
ax.text(2.3, 1.2, 'a + b')
ax.text(4.5, 2.08, add_vectors, color='fuchsia')
ax.set_facecolor('honeydew')
plt.show()
```

Output:

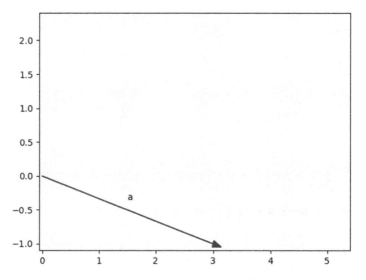

Figure 3-1. *Vector a from the origin (0, 0) to (3, -1)*

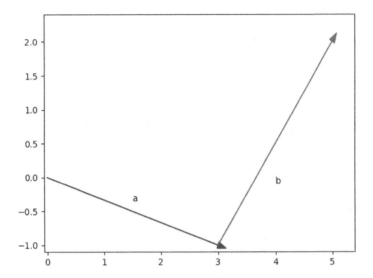

Figure 3-2. *Vector b from (3, -1) to (5, 2)*

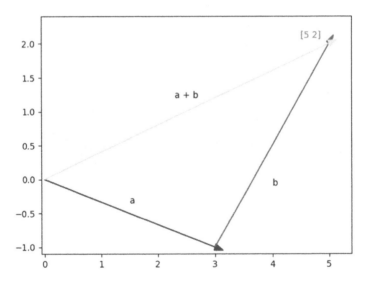

Figure 3-3. *Vector c from (0, 0) to (5, 2)*

The code begins by importing matplotlib and numpy libraries. Library matplotlib is a plotting library used for high quality visualization. Library numpy is the fundamental package for scientific computing. It is a wonderful library for working with vectors and matrices. The code continues with two functions–vector_add() and set_up(). Function vector_add() adds two vectors. Function set_up() sets up the figure for plotting. The main block begins by creating two vectors and adding them. The remainder of the code demonstrates graphically how vector addition works. First, it creates an axes() object with an arrow representing vector a beginning at origin (0, 0) and ending at (3, -1). It continues by adding text and a background color. Next, it creates a 2nd axes() object with the same arrow a, but adds arrow b (vector b) starting at (3, -1) and continuing to (2, 3). Finally, it creates a 3rd axes() object with the same arrows a and b, but adds arrow c (a + b) starting at (0, 0) and ending at (5, 2).

The 2nd example modifies the previous example by using subplots (Figure 3-4). Subplots divide a figure into an m × n grid for a different visualization experience.

```
import matplotlib.pyplot as plt, numpy as np

def vector_add(a, b):
    return np.add(a, b)

if __name__ == "__main__":
    v1, v2 = np.array([3, -1]), np.array([2, 3])
    add_vectors = vector_add(v1, v2)
    f, ax = plt.subplots(3)
    x, y = [0, 3], [0, -1]
    ax[0].set_xlim([-0.05, 3.1])
    ax[0].set_ylim([-1.1, 0.1])
    ax[0].scatter(x,y,s=1)
    ax[0].arrow(0, 0, 3, -1, head_width=0.1, head_length=0.07,
                fc='b', ec='b')
```

```
ax[0].text(1.5, -0.35, 'a')
ax[0].set_facecolor('honeydew')
x, y = ([0, 3, 5]), ([0, -1, 2])
ax[1].set_xlim([-0.05, 5.1])
ax[1].set_ylim([-1.2, 2.2])
ax[1].scatter(x,y,s=0.5)
ax[1].arrow(0, 0, 3, -1, head_width=0.2, head_length=0.1,
            fc='b', ec='b')
ax[1].arrow(3, -1, 2, 3, head_width=0.16, head_length=0.1,
            fc='crimson', ec='crimson')
ax[1].text(1.5, -0.35, 'a')
ax[1].text(4, -0.1, 'b')
ax[1].set_facecolor('honeydew')
x, y = ([0, 3, 5]), ([0, -1, 2])
ax[2].set_xlim([-0.05, 5.25])
ax[2].set_ylim([-1.2, 2.3])
ax[2].scatter(x,y,s=0.5)
ax[2].arrow(0, 0, 3, -1, head_width=0.15, head_length=0.1,
            fc='b', ec='b')
ax[2].arrow(3, -1, 2, 3, head_width=0.15, head_length=0.1,
            fc='crimson', ec='crimson')
ax[2].arrow(0, 0, 5, 2, head_width=0.1, head_length=0.1,
            fc='springgreen', ec='springgreen')
ax[2].text(1.5, -0.35, 'a')
ax[2].text(4, -0.1, 'b')
ax[2].text(2.3, 1.2, 'a + b')
ax[2].text(4.9, 1.4, add_vectors, color='fuchsia')
ax[2].set_facecolor('honeydew')
plt.tight_layout()
plt.show()
```

Output:

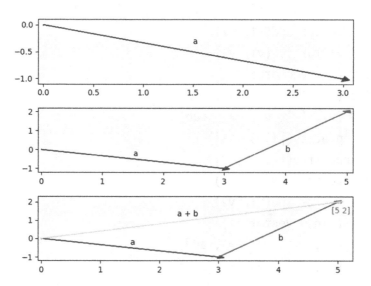

Figure 3-4. *Subplot Visualization of Vector Addition*

The code begins by importing matplotlib and numpy libraries. It continues with the same vector_add() function. The main block creates three subplots with plt.subplots(3) and assigns to f and ax, where f represents the figure and ax represents each subplot (ax[0], ax[1], and ax[2]). Instead of working with one figure, the code builds each subplot by indexing ax. The code uses plt.tight_layout() to automatically align each subplot.

The 3rd example adds vector subtraction. Subtracting two vectors is addition with the opposite (negation) of a vector. So, vector a minus vector b is the same as a + (-b). The code example demonstrates vector addition and subtraction for both 2- and 3-D vectors:

```python
import numpy as np

def vector_add(a, b):
    return np.add(a, b)

def vector_sub(a, b):
    return np.subtract(a, b)
```

```
if __name__ == "__main__":
    v1, v2 = np.array([3, -1]), np.array([2, 3])
    add = vector_add(v1, v2)
    sub = vector_sub(v1, v2)
    print ('2D vectors:')
    print (v1, '+', v2, '=', add)
    print (v1, '-', v2, '=', sub)
    v1 = np.array([1, 3, -5])
    v2 = np.array([2, -1, 3])
    add = vector_add(v1, v2)
    sub = vector_sub(v1, v2)
    print ('\n3D vectors:')
    print (v1, '+', v2, '=', add)
    print (v1, '-', v2, '=', sub)
```

Output:

```
2D vectors:
[ 3 -1] + [2 3] = [5 2]
[ 3 -1] - [2 3] = [ 1 -4]

3D vectors:
[ 1  3 -5] + [ 2 -1  3] = [ 3  2 -2]
[ 1  3 -5] - [ 2 -1  3] = [-1  4 -8]
```

The code begins by importing the numpy library. It continues with functions vector_add() and vector_subtract(), which add and subtract vectors respectively. The main block begins by creating two 2-D vectors, and adding and subtracting them. It continues by adding and subtracting two 3-D vectors. Any n-dimensional can be added and subtracted in the same manner.

Magnitude is measured by the distance formula. Magnitude of a single vector is measured from the origin $(0, 0)$ to the vector. Magnitude between two vectors is measured from the 1st vector to the 2nd vector. The distance formula is the square root of ((the 1st value from the 2nd vector minus the 1st value from the 1st vector squared) plus (the 2nd value from the 2nd vector minus the 2nd value from the 1st vector squared)).

Matrix Math

A matrix is an array of numbers. Many operations can be performed on a matrix such as addition, subtraction, negation, multiplication, and division. The dimension of a matrix is its size in number of rows and columns in that order. That is, a 2 × 3 matrix has two rows and three columns. Generally, an m × n matrix has m rows and n columns. An element is an entry in a matrix. Specifically, an element in row_i and $column_j$ of matrix A is denoted as $a_{i,j}$. Finally, a vector in a matrix is typically viewed as a column. So, a 2 × 3 matrix has three vectors (columns) each with two elements. This is a very important concept to understand when performing matrix multiplication and/or using matrices in data science algorithms.

The 1st code example creates a numpy matrix, multiplies it by a scalar, calculates means row- and column-wise, creates a numpy matrix from numpy arrays, and displays it by row and element:

```
import numpy as np

def mult_scalar(m, s):
    matrix = np.empty(m.shape)
    m_shape = m.shape
    for i, v in enumerate(range(m_shape[0])):
        result = [x * s for x in m[v]]
        x = np.array(result[0])
        matrix[i] = x
    return matrix

def display(m):
    s = np.shape(m)
    cols = s[1]
```

```
    for i, row in enumerate(m):
        print ('row', str(i) + ':', row, 'elements:', end=' ')
        for col in range(cols):
            print (row[col], end=' ')
        print ()

if __name__ == "__main__":
    v1, v2, v3 = [1, 7, -4], [2, -3, 10], [3, 5, 6]
    A = np.matrix([v1, v2, v3])
    print ('matrix A:\n', A)
    scalar = 0.5
    B = mult_scalar(A, scalar)
    print ('\nmatrix B:\n', B)
    mu_col = np.mean(A, axis=0, dtype=np.float64)
    print ('\nmean A (column-wise):\n', mu_col)
    mu_row = np.mean(A, axis=1, dtype=np.float64)
    print ('\nmean A (row-wise):\n', mu_row)
    print ('\nmatrix C:')
    C = np.array([[2, 14, -8], [4, -6, 20], [6, 10, 12]])
    print (C)
    print ('\ndisplay each row and element:')
    display(C)
```

Output:

```
matrix A:
 [[ 1  7 -4]
 [ 2 -3 10]
 [ 3  5  6]]

matrix B:
 [[ 0.5  3.5 -2. ]
 [ 1.  -1.5  5. ]
 [ 1.5  2.5  3. ]]

mean A (column-wise):
 [[ 2.   3.   4.]]

mean A (row-wise):
 [[ 1.33333333]
 [ 3.        ]
 [ 4.66666667]]

matrix C:
[[ 2 14 -8]
 [ 4 -6 20]
 [ 6 10 12]]

display each row and element:
row 0: [ 2 14 -8] elements: 2 14 -8
row 1: [ 4 -6 20] elements: 4 -6 20
row 2: [ 6 10 12] elements: 6 10 12
```

The code begins by importing numpy. It continues with two functions–mult_scalar() and display(). Function mult_scalar() multiplies a matrix by a scalar. Function display() displays a matrix by row and each element of a row. The main block creates three vectors and adds them to numpy matrix A. B is created by multiplying scalar 0.5 by A. Next, means for A are calculated by column and row. Finally, numpy matrix C is created from three numpy arrays and displayed by row and element.

The 2nd code example creates a numpy matrix A, sums its columns and rows, calculates the dot product of two vectors, and calculates the dot product of two matrices. Dot product multiplies two vectors to get magnitude that can be used to compute lengths of vectors and angles between vectors. Specifically, the dot product of two vectors a and b is $a_x \times b_x + a_y \times b_y$.

For matrix multiplication, dot product produces matrix C from two matrices A and B. However, two vectors cannot be multiplied when both are viewed as column matrices. To rectify this problem, transpose the 1st vector from A, turning it into a 1 × n row matrix so it can be multiplied by the 1st vector from B and summed. The product is now well defined because the product of a 1 × n matrix with an n × 1 matrix is a 1 × 1 matrix (a scalar). To get the dot product, repeat this process for the remaining vectors from A and B. Numpy includes a handy function that calculates dot product for you, which greatly simplifies matrix multiplication.

```python
import numpy as np

def sum_cols(matrix):
    return np.sum(matrix, axis=0)

def sum_rows(matrix):
    return np.sum(matrix, axis=1)

def dot(v, w):
    return np.dot(v, w)

if __name__ == "__main__":
    v1, v2, v3 = [1, 7, -4], [2, -3, 10], [3, 5, 6]
    A = np.matrix([v1, v2, v3])
    print ('matrix A:\n', A)
    v_cols = sum_cols(A)
    print ('\nsum A by column:\n', v_cols)
    v_rows = sum_rows(A)
    print ('\nsum A by row:\n', v_rows)
    dot_product = dot(v1, v2)
    print ('\nvector 1:', v1)
    print ('vector 2:', v2)
    print ('\ndot product v1 and v2:')
    print (dot_product)
```

```
v1, v2, v3 = [-2, 5, 4], [1, 2, 9], [10, -9, 3]
B = np.matrix([v1, v2, v3])
print ('\nmatrix B:\n', B)
C = A.dot(B)
print ('\nmatrix C (dot product A and B):\n', C)
print ('\nC by row:')
for i, row in enumerate(C):
    print ('row', str(i) + ': ', end='')
    for v in np.nditer(row):
        print (v, end=' ')
    print()
```

Output:

```
matrix A:
 [[ 1  7 -4]
 [ 2 -3 10]
 [ 3  5  6]]

sum A by column:
 [[ 6  9 12]]

sum A by row:
 [[ 4]
 [ 9]
 [14]]

vector 1: [1, 7, -4]
vector 2: [2, -3, 10]

dot product v1 and v2:
-59

matrix B:
 [[-2  5  4]
 [ 1  2  9]
 [10 -9  3]]

matrix C (dot product A and B):
 [[-35  55  55]
 [ 93 -86  11]
 [ 59 -29  75]]

C by row:
row 0: -35 55 55
row 1: 93 -86 11
row 2: 59 -29 75
```

The code begins by importing numpy. It continues with three functions–sum_cols(), sum_rows(), and dot(). Function sum_cols() sums each column and returns a row with these values. Function sum_rows() sums each row and returns a column with these values. Function dot() calculates the dot product. The main block begins by creating three vectors that are then used to create matrix A. Columns and rows are summed for A. Dot product is then calculated for two vectors (v1 and v2). Next, three new vectors are created that are then used to create matrix B. Matrix C is created by calculating the dot product for A and B. Finally, each row of C is displayed.

The 3rd code example illuminates a realistic scenario. Suppose a company sells three types of pies–beef, chicken, and vegetable. Beef pies cost $3 each, chicken pies cost $4 dollars each, and vegetable pies cost $2 dollars each. The vector representation for pie cost is [3, 4, 2]. You also know sales by pie for Monday through Thursday. Beef sales are 13 for Monday, 9 for Tuesday, 7 for Wednesday, and 15 for Thursday. The vector for beef sales is thereby [13, 9, 7, 15]. Using the same logic, the vectors for chicken sales are [8, 7, 4, 6] and [6, 4, 0, 3], respectively. The goal is to calculate total sales for four days (Monday-Thursday).

```
import numpy as np

def dot(v, w):
    return np.dot(v, w)

def display(m):
    for i, row in enumerate(m):
        print ('total sales by day:\n', end='')
        for v in np.nditer(row):
            print (v, end=' ')
        print()
```

```python
if __name__ == "__main__":
    a = [3, 4, 2]
    A = np.matrix([a])
    print ('cost matrix A:\n', A)
    v1, v2, v3 = [13, 9, 7, 15], [8, 7, 4, 6], [6, 4, 0, 3]
    B = np.matrix([v1, v2, v3])
    print ('\ndaily sales by item matrix B:\n', B)
    C = A.dot(B)
    print ('\ndot product matrix C:\n', C, '\n')
    display(C)
```

Output:

```
cost matrix A:
[[3 4 2]]

daily sales by item matrix B:
[[13  9  7 15]
 [ 8  7  4  6]
 [ 6  4  0  3]]

dot product matrix C:
[[83 63 37 75]]

total sales by day:
83 63 37 75
```

The code begins by importing numpy. It continues with function dot() that calculates the dot product, and function display() that displays the elements of a matrix, row by row. The main block begins by creating a vector that holds the cost of each type of pie. It continues by converting the vector into matrix A. Next, three vectors are created that represent sales for each type of pie for Monday through Friday. The code continues by converting the three vectors into matrix B. Matrix C is created by finding the dot product of A and B. This scenario demonstrates how dot product can be used for solving business problems.

The 4th code example calculates the magnitude (distance) and direction (angle) with a single vector and between two vectors:

```
import math, numpy as np

def sqrt_sum_squares(ls):
    return math.sqrt(sum(map(lambda x:x*x,ls)))

def mag(v):
    return np.linalg.norm(v)

def a_tang(v):
    return math.degrees(math.atan(v[1]/v[0]))

def dist(v, w):
    return math.sqrt((((w[0]-v[0])** 2) + ((w[1]-v[1])** 2))

def mags(v, w):
    return np.linalg.norm(v - w)

def a_tangs(v, w):
    val = (w[1] - v[1]) / (w[0] - v[0])
    return math.degrees(math.atan(val))

if __name__ == "__main__":
    v = np.array([3, 4])
    print ('single vector', str(v) + ':')
    print ('magnitude:', sqrt_sum_squares(v))
    print ('NumPY magnitude:', mag(v))
    print ('direction:', round(a_tang(v)), 'degrees\n')
    v1, v2 = np.array([2, 3]), np.array([5, 8])
    print ('two vectors', str(v1) + ' and ' + str(v2) + ':')
    print ('magnitude', round(dist(v1, v2),2))
    print ('NumPY magnitude:', round(mags(v1, v2),2))
    print ('direction:', round(a_tangs(v1, v2)), 'degrees\n')
    v1, v2 = np.array([0, 0]), np.array([3, 4])
```

```
print ('use origin (0,0) as 1st vector:')
print ('"two vectors', str(v1) + ' and ' + str(v2) + '"')
print ('magnitude:', round(mags(v1, v2),2))
print ('direction:', round(a_tangs(v1, v2)), 'degrees')
```

Output:

```
single vector [3 4]:
magnitude: 5.0
NumPY magnitude: 5.0
direction: 53 degrees

two vectors [2 3] and [5 8]:
magnitude 5.83
NumPY magnitude: 5.83
direction: 59 degrees

use origin (0,0) as 1st vector:
"two vectors [0 0] and [3 4]"
magnitude: 5.0
direction: 53 degrees
```

The code begins by importing math and numpy libraries. It continues with six functions. Function sqrt_sum_squares() calculates magnitude for one vector from scratch. Function mag() does the same but uses numpy. Function a_tang() calculates the arctangent of a vector, which is the direction (angle) of a vector from the origin (0,0). Function dist() calculates magnitude between two vectors from scratch. Function mags() does the same but uses numpy. Function a_tangs() calculates the arctangent of two vectors. The main block creates a vector, calculates magnitude and direction, and displays. Next, magnitude and direction are calculated and displayed for two vectors. Finally, magnitude and direction for a single vector are calculated using the two vector formulas. This is accomplished by using the origin (0,0) as the 1st vector. So, functions that calculate magnitude and direction for a single vector are not needed, because any single vector always begins from the origin (0,0). Therefore, a vector is simply a point in space measured either from the origin (0,0) or in relation to another vector by magnitude and direction.

Basic Matrix Transformations

The 1st code example introduces the identity matrix, which is a square matrix with ones on the main diagonal and zeros elsewhere. The product of matrix A and its identity matrix is A, which is important mathematically because the identity property of multiplication states that any number multiplied by 1 is equal to itself.

```python
import numpy as np

def slice_row(M, i):
    return M[i,:]

def slice_col(M, j):
    return M[:, j]

def to_int(M):
    return M.astype(np.int64)

if __name__ == "__main__":
    A = [[1, 9, 3, 6, 7],
         [4, 8, 6, 2, 1],
         [9, 8, 7, 1, 2],
         [1, 1, 9, 2, 4],
         [9, 1, 1, 3, 5]]
    A = np.matrix(A)
    print ('A:\n', A)
    print ('\n1st row: ', slice_row(A, 0))
    print ('\n3rd column:\n', slice_col(A, 2))
    shapeA = np.shape(A)
    I = np.identity(np.shape(A)[0])
    I = to_int(I)
    print ('\nI:\n', I)
    dot_product = np.dot(A, I)
```

```
print ('\nA * I = A:\n', dot_product)
print ('\nA\':\n', A.I)
A_by_Ainv = np.round(np.dot(A, A.I), decimals=0, out=None)
A_by_Ainv = to_int(A_by_Ainv)
print ('\nA * A\':\n', A_by_Ainv)
```

Output:

```
A:
 [[1 9 3 6 7]
 [4 8 6 2 1]
 [9 8 7 1 2]
 [1 1 9 2 4]
 [9 1 1 3 5]]

1st row:  [[1 9 3 6 7]]

3rd column:
 [[3]
 [6]
 [7]
 [9]
 [1]]

I:
 [[1 0 0 0 0]
 [0 1 0 0 0]
 [0 0 1 0 0]
 [0 0 0 1 0]
 [0 0 0 0 1]]

A * I = A:
 [[1 9 3 6 7]
 [4 8 6 2 1]
 [9 8 7 1 2]
 [1 1 9 2 4]
 [9 1 1 3 5]]

A':
 [[ -6.12745098e-02   6.90144479e-02  -8.77192982e-03  -2.97987616e-02
    9.93292054e-02]
 [  8.82352941e-02  -1.09907121e-01   1.57894737e-01  -6.65634675e-02
   -1.11455108e-01]
 [ -5.63725490e-02   9.50722394e-02  -4.38596491e-02   1.01006192e-01
   -3.35397317e-03]
 [ -1.61764706e-01   8.24303406e-01  -6.84210526e-01  -7.73993808e-04
    3.35913313e-01]
 [  2.00980392e-01  -6.15841073e-01   4.03508772e-01   4.72136223e-02
   -1.57378741e-01]]

A * A':
 [[1 0 0 0 0]
 [0 1 0 0 0]
 [0 0 1 0 0]
 [0 0 0 1 0]
 [0 0 0 0 1]]
```

The code begins by importing numpy. It continues with three functions. Function slice_row() slices a row from a matrix. Function slice_col() slices a column from a matrix. Function to_int() converts matrix

85

elements to integers. The main block begins by creating matrix A.
It continues by creating the identity matrix for A. Finally, it creates the
identity matrix for A by using the dot product of A with A' (inverse of A).

The 2nd code example converts a list of lists into a numpy matrix and
traverses it:

```python
import numpy as np

if __name__ == "__main__":
    data = [
        [41, 72, 180], [27, 66, 140],
        [18, 59, 101], [57, 72, 160],
        [21, 59, 112], [29, 77, 250],
        [55, 60, 120], [28, 72, 110],
        [19, 59, 99], [32, 68, 125],
        [31, 79, 322], [36, 69, 111]
        ]
    A = np.matrix(data)
    print ('manual traversal:')
    for p in range(A.shape[0]):
        for q in range(A.shape[1]):
            print (A[p,q], end=' ')
        print ()
```

Output:

```
manual traversal:
41 72 180
27 66 140
18 59 101
57 72 160
21 59 112
29 77 250
55 60 120
28 72 110
19 59 99
32 68 125
31 79 322
36 69 111
```

The code begins by importing numpy. The main block begins by creating a list of lists, converting it into numpy matrix A, and traversing A. Although I have demonstrated several methods for traversing a numpy matrix, this is my favorite method.

The 3rd code example converts a list of lists into numpy matrix A. It then slices and dices A:

```python
import numpy as np

if __name__ == "__main__":
    points_3D_space = [
        [0, 0, 0],
        [1, 2, 3],
        [2, 2, 2],
        [9, 9, 9] ]
    A = np.matrix(points_3D_space)
    print ('slice entire A:')
    print (A[:])
    print ('\nslice 2nd column:')
    print (A[0:4, 1])
    print ('\nslice 2nd column (alt method):')
    print (A[:, 1])
    print ('\nslice 2nd & 3rd value 3rd column:')
    print (A[1:3, 2])
    print ('\nslice last row:')
    print (A[-1])
    print ('\nslice last row (alt method):')
    print (A[3])
    print ('\nslice 1st row:')
    print (A[0, :])
    print ('\nslice 2nd row; 2nd & 3rd value:')
    print (A[1, 1:3])
```

Output:

```
slice entire A:
[[0 0 0]
 [1 2 3]
 [2 2 2]
 [9 9 9]]

slice 2nd column:
[[0]
 [2]
 [2]
 [9]]

slice 2nd column (alt method):
[[0]
 [2]
 [2]
 [9]]

slice 2nd & 3rd value 3rd column:
[[3]
 [2]]

slice last row:
[[9 9 9]]

slice last row (alt method):
[[9 9 9]]

slice 1st row:
[[0 0 0]]

slice 2nd row; 2nd & 3rd value:
[[2 3]]
```

The code begins by importing numpy. The main block begins by creating a list of lists and converting it into numpy matrix A. The code continues by slicing and dicing the matrix.

Pandas Matrix Applications

The pandas library provides high-performance, easy-to-use data structure and analysis tools. The most commonly used pandas object is a DataFrame (df). A df is a 2-D structure with labeled axes (row and column) of potentially different types. Math operations align on both row and column labels. A df can be conceptualized by column or row. To view by column, use axis = 0 or axis = 'index'. To view by row, use axis = 1 or axis = 'columns'. This may seem counterintuitive when working with rows, but this is the way pandas implemented this feature.

A pandas df is much easier to work with than a numpy matrix, but it is also less efficient. That is, it takes a lot more resources to process a pandas df. The numpy library is optimized for processing large amounts of data and numerical calculations.

The 1st example creates a list of lists, places it into a pandas df, and displays some data:

```python
import pandas as pd

if __name__ == "__main__":
    data = [
        [41, 72, 180], [27, 66, 140],
        [18, 59, 101], [57, 72, 160],
        [21, 59, 112], [29, 77, 250],
        [55, 60, 120], [28, 72, 110],
        [19, 59, 99], [32, 68, 125],
        [31, 79, 322], [36, 69, 111]
        ]
    headers = ['age', 'height', 'weight']
    df = pd.DataFrame(data, columns=headers)
    n = 3
    print ('First', n, '"df" rows:\n', df.head(n))
    print ('\nFirst "df" row:')
    print (df[0:1])
    print ('\nRows 2 through 4')
    print (df[2:5])
    print ('\nFirst', n, 'rows "age" column')
    print (df[['age']].head(n))
    print ('\nLast', n, 'rows "weight" and "age" columns')
    print (df[['weight', 'age']].tail(n))
```

```
print ('\nRows 3 through 6 "weight" and "age" columns')
print (df.ix[3:6, ['weight', 'age']])
```

Output:

```
First 3 "df" rows:
    age  height  weight
0    41      72     180
1    27      66     140
2    18      59     101

First "df" row:
    age  height  weight
0    41      72     180

Rows 2 through 4
    age  height  weight
2    18      59     101
3    57      72     160
4    21      59     112

First 3 rows "age" column
    age
0    41
1    27
2    18

Last 3 rows "weight" and "age" columns
     weight  age
9       125   32
10      322   31
11      111   36

Rows 3 through 6 "weight" and "age" columns
    weight  age
3      160   57
4      112   21
5      250   29
6      120   55
```

The code begins by importing pandas. The main block begins by creating a list of lists and adding it to a pandas df. It is a good idea to create your own headers as we do here. Method head() and tail() automatically display the 1st five records and last five records respectively unless a value is included. In this case, we display the 1st and last three records. Using head() and tail() are very useful, especially with a large df. Notice how easy it is to slice and dice the df. Also, notice how easy it is to display column data of your choice.

The 2nd example creates a list of lists, places it into numpy matrix A, and puts A into a pandas df. This ability is very important because it shows how easy it is to create a df from a numpy matrix. So, you can be working with numpy matrices for precision and performance, and then convert to pandas for slicing, dicing, and other operations.

```
import pandas as pd, numpy as np

if __name__ == "__main__":
    data = [
        [41, 72, 180], [27, 66, 140],
        [18, 59, 101], [57, 72, 160],
        [21, 59, 112], [29, 77, 250],
        [55, 60, 120], [28, 72, 110],
        [19, 59, 99], [32, 68, 125],
        [31, 79, 322], [36, 69, 111]
        ]
    A = np.matrix(data)
    headers = ['age', 'height', 'weight']
    df = pd.DataFrame(A, columns=headers)
    print ('Entire "df":')
    print (df, '\n')
    print ('Sliced by "age" and "height":')
    print (df[['age', 'height']])
```

Output:

```
Entire "df":
    age  height  weight
0    41      72     180
1    27      66     140
2    18      59     101
3    57      72     160
4    21      59     112
5    29      77     250
6    55      60     120
7    28      72     110
8    19      59      99
9    32      68     125
10   31      79     322
11   36      69     111

Sliced by "age" and "height":
    age  height
0    41      72
1    27      66
2    18      59
3    57      72
4    21      59
5    29      77
6    55      60
7    28      72
8    19      59
9    32      68
10   31      79
11   36      69
```

The code begins by importing pandas and numpy. The main block begins by creating a list of lists, converting it to numpy matrix A, and then adding A to a pandas df.

The 3rd example creates a list of lists, places it into a list of dictionary elements, and puts it into a pandas df. This ability is also very important because dictionaries are very efficient data structures when working with data science applications.

```python
import pandas as pd

if __name__ == "__main__":
    data = [
        [41, 72, 180], [27, 66, 140],
        [18, 59, 101], [57, 72, 160],
        [21, 59, 112], [29, 77, 250],
        [55, 60, 120], [28, 72, 110],
        [19, 59, 99], [32, 68, 125],
        [31, 79, 322], [36, 69, 111]
        ]
    d = {}
    dls = []
    key = ['age', 'height', 'weight']
    for row in data:
        for i, num in enumerate(row):
            d[key[i]] = num
        dls.append(d)
        d = {}
    df = pd.DataFrame(dls)
    print ('dict elements from list:')
    for row in dls:
        print (row)
    print ('\nheight from 1st dict element is:', end=' ')
    print (dls[0]['height'])
    print ('\n"df" converted from dict list:\n', df)
    print ('\nheight 1st df element:\n', df[['height']].head(1))
```

Output:

```
dict elements from list:
{'age': 41, 'height': 72, 'weight': 180}
{'age': 27, 'height': 66, 'weight': 140}
{'age': 18, 'height': 59, 'weight': 101}
{'age': 57, 'height': 72, 'weight': 160}
{'age': 21, 'height': 59, 'weight': 112}
{'age': 29, 'height': 77, 'weight': 250}
{'age': 55, 'height': 60, 'weight': 120}
{'age': 28, 'height': 72, 'weight': 110}
{'age': 19, 'height': 59, 'weight': 99}
{'age': 32, 'height': 68, 'weight': 125}
{'age': 31, 'height': 79, 'weight': 322}
{'age': 36, 'height': 69, 'weight': 111}

height from 1st dict element is: 72

"df" converted from dict list:
    age  height  weight
0   41      72     180
1   27      66     140
2   18      59     101
3   57      72     160
4   21      59     112
5   29      77     250
6   55      60     120
7   28      72     110
8   19      59      99
9   32      68     125
10  31      79     322
11  36      69     111

height 1st df element:
    height
0       72
```

The 4th code example creates two lists of lists–data and scores. The data list holds ages, heights, and weights for 12 athletes. The scores list holds three exam scores for 12 students. The data list is put directly into df1, and the scores list is put directly into df2. Averages are computed and displayed.

```
import pandas as pd, numpy as np

if __name__ == "__main__":
    data = [
        [41, 72, 180], [27, 66, 140],
        [18, 59, 101], [57, 72, 160],
        [21, 59, 112], [29, 77, 250],
```

```
    [55, 60, 120], [28, 72, 110],
    [19, 59, 99], [32, 68, 125],
    [31, 79, 322], [36, 69, 111]
    ]
scores = [
    [99, 90, 88], [77, 66, 81], [78, 77, 83],
    [75, 72, 79], [88, 77, 93], [88, 77, 94],
    [100, 99, 93], [94, 74, 90], [98, 97, 99],
    [73, 68, 77], [55, 50, 68], [36, 77, 90]
    ]
n = 3
key1 = ['age', 'height', 'weight']
df1 = pd.DataFrame(data, columns=key1)
print ('df1 slice:\n', df1.head(n))
avg_cols = df1.apply(np.mean, axis=0)
print ('\naverage by columns:')
print (avg_cols)
avg_wt = df1[['weight']].apply(np.mean, axis='index')
print ('\naverage weight')
print (avg_wt)
key2 = ['exam1', 'exam2', 'exam3']
df2 = pd.DataFrame(scores, columns=key2)
print ('\ndf2 slice:\n', df2.head(n))
avg_scores = df2.apply(np.mean, axis=1)
print ('\naverage scores for 1st', n, 'students (rows):')
print (avg_scores.head(n))
avg_slice = df2[['exam1','exam3']].apply(np.mean,
axis='columns')
print ('\naverage "exam1" & "exam3" 1st', n, 'students
(rows):')
print (avg_slice[0:n])
```

Output:

```
df1 slice:
    age  height  weight
0    41      72     180
1    27      66     140
2    18      59     101

average by columns:
age          32.833333
height       67.666667
weight      152.500000
dtype: float64

average weight
weight      152.5
dtype: float64

df2 slice:
    exam1  exam2  exam3
0      99     90     88
1      77     66     81
2      78     77     83

average scores for 1st 3 students (rows):
0    92.333333
1    74.666667
2    79.333333
dtype: float64

average "exam1" & "exam3" 1st 3 students (rows):
0    93.5
1    79.0
2    80.5
dtype: float64
```

The code begins by importing pandas and numpy. The main block creates the data and scores lists and puts them in df1 and df2, respectively. With df1 (data), we average by column because our goal is to return the average age, height, and weight for all athletes. With df2 (scores), we average by row because our goal is to return the average overall exam score for each student. We could average by column for df2 if the goal is to calculate the average overall score for one of the exams. Try this if you wish.

CHAPTER 4

Gradient Descent

Gradient descent (GD) is an algorithm that minimizes (or maximizes) functions. To apply, start at an initial set of a function's parameter values and iteratively move toward a set of parameter values that minimize the function. Iterative minimization is achieved using calculus by taking steps in the negative direction of the function's gradient. GD is important because optimization is a big part of machine learning. Also, GD is easy to implement, generic, and efficient (fast).

Simple Function Minimization (and Maximization)

GD is a 1st order iterative optimization algorithm for finding the minimum of a function f. A function can be denoted as f or f(x). Simply, GD finds the minimum error by minimizing (or maximizing) a cost function. A cost function is something that you want to minimize.

Let's begin with a minimization example. To find the local minimum of f, take steps proportional to the negative of the gradient of f at the current point. The gradient is the derivative (rate of change) of f. The only weakness of GD is that it finds the local minimum rather than the minimum for the whole function.

© David Paper 2018
D. Paper, *Data Science Fundamentals for Python and MongoDB*,
https://doi.org/10.1007/978-1-4842-3597-3_4

The power rule is used to differentiate functions of the form $f(x) = x^r$:

$$\frac{d}{dx}x^n = nx^{n-1}$$

So, the derivative of x^n equals nx^{n-1}. Simply, the derivative is the product of the exponent times x with the exponent reduced by 1. To minimize $f(x) = x^4 - 3x^3 + 2$ find the derivative, which is $f'(x) = 4x^3 - 9x^2$. So, the 1st step is always to find the derivative $f'(x)$. The 2nd step is to plot the original function to get an idea of its shape. The 3rd step is to run GD. The 4th step is to plot the local minimum.

The 1st example finds the local minimum of $f(x)$ and displays $f(x)$, $f'(x)$, and minimum in the subplot as seen in Figure 4-1:

```
import matplotlib.pyplot as plt, numpy as np

def f(x):
    return x**4 - 3 * x**3 + 2

def df(x):
    return 4 * x**3 - 9 * x**2

if __name__ == "__main__":
    x = np.arange(-5, 5, 0.2)
    y, y_dx = f(x), df(x)
    f, axarr = plt.subplots(3, sharex=True)
    axarr[0].plot(x, y, color='mediumspringgreen')
    axarr[0].set_xlabel('x')
    axarr[0].set_ylabel('f(x)')
    axarr[0].set_title('f(x)')
    axarr[1].plot(x, y_dx, color='coral')
    axarr[1].set_xlabel('x')
    axarr[1].set_ylabel('dy/dx(x)')
    axarr[1].set_title('derivative of f(x)')
    axarr[2].set_xlabel('x')
    axarr[2].set_ylabel('GD')
```

```python
axarr[2].set_title('local minimum')
iterations, cur_x, gamma, precision = 0, 6, 0.01, 0.00001
previous_step_size = cur_x
while previous_step_size > precision:
    prev_x = cur_x
    cur_x += -gamma * df(prev_x)
    previous_step_size = abs(cur_x - prev_x)
    iterations += 1
    axarr[2].plot(prev_x, cur_x, "o")
f.subplots_adjust(hspace=0.3)
f.tight_layout()
plt.show()
print ('minimum:', cur_x, '\niterations:', iterations)
```

Output:

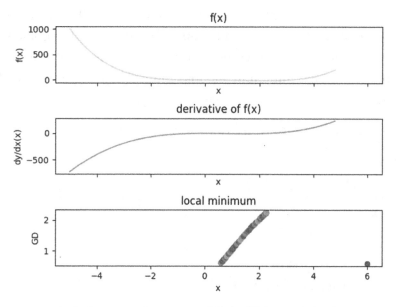

Figure 4-1. *Subplot visualization of f(x), f'(x), and the local minimum*

The code example begins by importing matplotlib and numpy. It continues with function f(x) used to plot the original function and function df(x) used to plot the derivative. The main block begins by creating values for f(x). It continues by creating a subplot. GD begins by initializing variables. Variable cur_x is the starting point for the simulation. Variable gamma is the step size. Variable precision is the tolerance. Smaller tolerance translates into more precision, but requires more iterations (resources). The simulation continues until previous_step_size is greater than precision. Each iteration multiplies -gamma (step_size) by the gradient (derivative) at the current point to move it to the local minimum. Variable previous_step_size is then assigned the difference between cur_x and prev_x. Each point is plotted. The minimum for f(x) solving for x is approximately 2.25. I know this result is correct because I calculated it by hand. Check out `http://www.dummies.com/education/math/calculus/how-to-find-local-extrema-with-the-first-derivative-test/` for a nice lesson on how to calculate by hand.

The 2nd example finds the local minimum and maximum of $f(x) = x^3 - 6x^2 + 9x + 15$. First find f'(x), which is $3x^2 - 12x + 9$. Next, find the local minimum, plot, local maximum, and plot. I don't use a subplot in this case because the visualization is not as rich. That is, it is much easier to see the approximate local minimum and maximum by looking at a plot of f(x), and easier to see how the GD process works its magic.

```
import matplotlib.pyplot as plt, numpy as np

def f(x):
    return x**3 - 6 * x**2 + 9 * x + 15

def df(x):
    return 3 * x**2 - 12 * x + 9

if __name__ == "__main__":
    x = np.arange(-0.5, 5, 0.2)
    y = f(x)
    plt.figure('f(x)')
```

```python
plt.xlabel('x')
plt.ylabel('f(x)')
plt.title('f(x)')
plt.plot(x, y, color='blueviolet')
plt.figure('local minimum')
plt.xlabel('x')
plt.ylabel('GD')
plt.title('local minimum')
iterations, cur_x, gamma, precision = 0, 6, 0.01, 0.00001
previous_step_size = cur_x
while previous_step_size > precision:
    prev_x = cur_x
    cur_x += -gamma * df(prev_x)
    previous_step_size = abs(cur_x - prev_x)
    iterations += 1
    plt.plot(prev_x, cur_x, "o")
local_min = cur_x
print ('minimum:', local_min, 'iterations:', iterations)
plt.figure('local maximum')
plt.xlabel('x')
plt.ylabel('GD')
plt.title('local maximum')
iterations, cur_x, gamma, precision = 0, 0.5, 0.01, 0.00001
previous_step_size = cur_x
while previous_step_size > precision:
    prev_x = cur_x
    cur_x += -gamma * -df(prev_x)
    previous_step_size = abs(cur_x - prev_x)
    iterations += 1
    plt.plot(prev_x, cur_x, "o")
local_max = cur_x
print ('maximum:', local_max, 'iterations:', iterations)
plt.show()
```

Output:

```
minimum: 3.0001526323101704 iterations: 144
maximum: 0.9998475518984531 iterations: 127
```

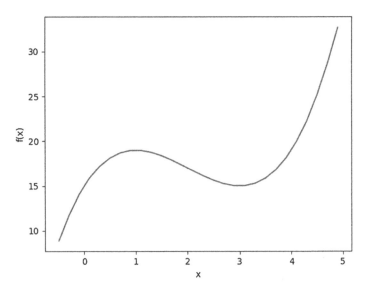

Figure 4-2. *Function f(x)*

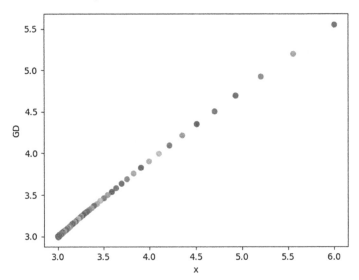

Figure 4-3. *Local minimum for function f(x)*

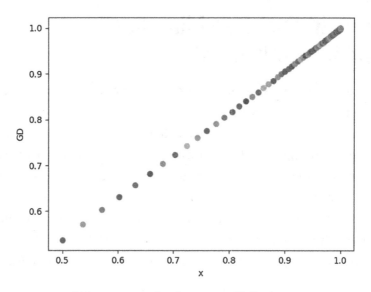

Figure 4-4. *Local maximum for function f(x)*

The code begins by importing matplotlib and numpy libraries. It continues with functions f(x) and df(x), which represent the original function and its derivative algorithmically. The main block begins by creating data for f(x) and plotting it. It continues by finding the local minimum and maximum, and plotting them. Notice the cur_x (the beginning point) for local minimum is 6, while it is 0.5 for local maximum. This is where data science is more of an art than a science, because I found these points by trial and error. Also notice that GD for the local maximum is the negation of the derivative. Again, I know that the results are correct because I calculated both local minimum and maximum by hand. The main reason that I used separate plots rather than a subplot for this example is to demonstrate why it is so important to plot f(x). Just by looking at the plot, you can tell that the local maximum of x for f(x) is close to one, and the local minimum of x for f(x) is close to 3. In addition, you can see that the function has an overall maximum that is greater than 1 from this plot. Figures 4-2, 4-3, and 4-4 provide the visualizations.

Sigmoid Function Minimization (and Maximization)

A sigmoid function is a mathematical function with an S-shaped or sigmoid curve. It is very important in data science for several reasons. First, it is easily differentiable with respect to network parameters, which are pivotal in training neural networks. Second, the cumulative distribution functions for many common probability distributions are sigmoidal. Third, many natural processes (e.g., complex learning curves) follow a sigmoidal curve over time. So, a sigmoid function is often used if no specific mathematical model is available.

The 1st example finds the local minimum of the sigmoid function:

```python
import matplotlib.pyplot as plt, numpy as np

def sigmoid(x):
    return 1 / (1 + np.exp(-x))

def df(x):
    return x * (1-x)

if __name__ == "__main__":
    x = np.arange(-10., 10., 0.2)
    y, y_dx = sigmoid(x), df(x)
    f, axarr = plt.subplots(3, sharex=True)
    axarr[0].plot(x, y, color='lime')
    axarr[0].set_xlabel('x')
    axarr[0].set_ylabel('f(x)')
    axarr[0].set_title('Sigmoid Function')
    axarr[1].plot(x, y_dx, color='coral')
    axarr[1].set_xlabel('x')
```

```python
axarr[1].set_ylabel('dy/dx(x)')
axarr[1].set_title('Derivative of f(x)')
axarr[2].set_xlabel('x')
axarr[2].set_ylabel('GD')
axarr[2].set_title('local minimum')
iterations, cur_x, gamma, precision = 0, 0.01, 0.01, 0.00001
previous_step_size = cur_x
while previous_step_size > precision:
    prev_x = cur_x
    cur_x += -gamma * df(prev_x)
    previous_step_size = abs(cur_x - prev_x)
    iterations += 1
    plt.plot(prev_x, cur_x, "o")
f.subplots_adjust(hspace=0.3)
f.tight_layout()
print ('minimum:', cur_x, '\niterations:', iterations)
plt.show()
```

Output:

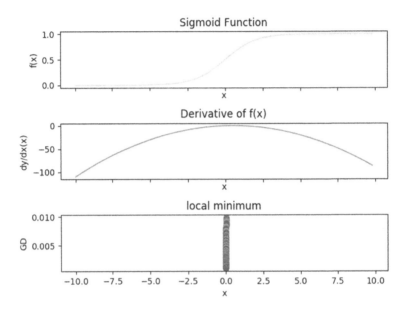

Figure 4-5. *Subplot of f(x), f'(x), and local minimum*

The code begins by importing matplotlib and numpy. It continues with functions sigmoid(x) and df(x), which represent the sigmoid function and its derivative algorithmically. The main block begins by creating data for f(x) and f'(x). It continues by creating subplots for f(x), f'(x), and the local minimum. In this case, using subplots was fine for visualization. It is easy to see from the f(x) and f'(x) plots (Figure 4-5) that the local minimum is close to 0. Next, the code runs GD to find the local minimum and plots it.

Again, the starting point for GD, cur_x, was found by trial and error. If you start cur_x further from the local minimum (you can estimate this by looking at the subplot of f'(x)), the number of iterations increases because it takes longer for the GD algorithm to converge on the local minimum. As expected, the local minimum is approximately 0.

The 2nd example finds the local maximum of the sigmoid function:

```python
import matplotlib.pyplot as plt, numpy as np

def sigmoid(x):
    return 1 / (1 + np.exp(-x))

def df(x):
    return x * (1-x)

if __name__ == "__main__":
    x = np.arange(-10., 10., 0.2)
    y, y_dx = sigmoid(x), df(x)
    f, axarr = plt.subplots(3, sharex=True)
    axarr[0].plot(x, y, color='lime')
    axarr[0].set_xlabel('x')
    axarr[0].set_ylabel('f(x)')
    axarr[0].set_title('Sigmoid Function')
    axarr[1].plot(x, y_dx, color='coral')
    axarr[1].set_xlabel('x')
    axarr[1].set_ylabel('dy/dx(x)')
    axarr[1].set_title('Derivative of f(x)')
    axarr[2].set_xlabel('x')
    axarr[2].set_ylabel('GD')
    axarr[2].set_title('local maximum')
    iterations, cur_x, gamma, precision = 0, 0.01, 0.01, 0.00001
    previous_step_size = cur_x
```

```
while previous_step_size > precision:
    prev_x = cur_x
    cur_x += -gamma * -df(prev_x)
    previous_step_size = abs(cur_x - prev_x)
    iterations += 1
    plt.plot(prev_x, cur_x, "o")
f.subplots_adjust(hspace=0.3)
f.tight_layout()
print ('maximum:', cur_x, '\niterations:', iterations)
plt.show()
```

Output:

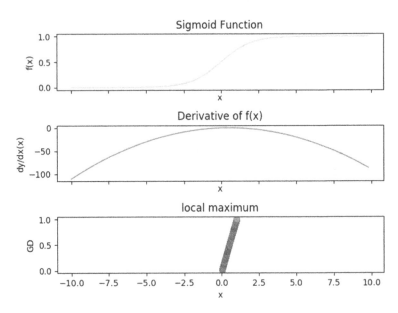

Figure 4-6. *Subplot of f(x), f'(x), and local maximum*

The code begins by importing matplotlib and numpy. It continues with functions sigmoid(x) and df(x), which represent the sigmoid function and its derivative algorithmically. The main block begins by creating data for f(x) and f'(x). It continues by creating subplots for f(x), f'(x), and the local maximum (Figure 4-6). It is easy to see from the f(x) plot that the local maximum is close to 1. Next, the code runs GD to find the local maximum and plots it. Again, the starting point for GD, cur_x, was found by trial and error. If you start cur_x further from the local maximum (you can estimate this by looking at the subplot of f(x)), the number of iterations increases because it takes longer for the GD algorithm to converge on the local maximum. As expected, the local maximum is approximately 1.

Euclidean Distance Minimization Controlling for Step Size

Euclidean distance is the ordinary straight-line distance between two points in Euclidean space. With this distance, Euclidean space becomes a metric space. The associated norm is the Euclidean norm (EN). The EN assigns each vector the length of its arrow. So, EN is really just the magnitude of a vector. A vector space on which a norm is defined is the normed vector space.

To find the local minimum of f(x) in three-dimensional (3-D) space, the 1st step is to find the minimum for all 3-D vectors. The 2nd step is to create a random 3-D vector [x, y, z]. The 3rd step is to pick a random starting point, and then take tiny steps in the opposite direction of the gradient f'(x) until a point is reached where the gradient is very small. Each tiny step (from the current vector to the next vector) is measured with the ED metric. The ED metric is the distance between two points in Euclidean space. The metric is required because we need to know how to move for each tiny step. So, the ED metric supplements GD to find the local minimum in 3-D space.

The code example finds the local minimum of the sigmoid function in 3-D space:

```python
import matplotlib.pyplot as plt
from mpl_toolkits.mplot3d import Axes3D
import random, numpy as np
from scipy.spatial import distance

def step(v, direction, step_size):
    return [v_i + step_size * direction_i
            for v_i, direction_i in zip(v, direction)]

def sigmoid_gradient(v):
    return [v_i * (1-v_i) for v_i in v]

def mod_vector(v):
    for i, v_i in enumerate(v):
        if v_i == float("inf") or v_i == float("-inf"):
            v[i] = random.randint(-1, 1)
    return v

if __name__ == "__main__":
    v = [random.randint(-10, 10) for i in range(3)]
    tolerance = 0.0000001
    iterations = 1
    fig = plt.figure('Euclidean')
    ax = fig.add_subplot(111, projection='3d')
    while True:
        gradient = sigmoid_gradient(v)
        next_v = step(v, gradient, -0.01)
        xs = gradient[0]
        ys = gradient[1]
        zs = gradient[2]
        ax.scatter(xs, ys, zs, c='lime', marker='o')
```

```
        v = mod_vector(v)
        next_v = mod_vector(next_v)
        test_v = distance.euclidean(v, next_v)
        if test_v < tolerance:
            break
        v = next_v
        iterations += 1
print ('minimum:', test_v, '\niterations:', iterations)
ax.set_xlabel('X axis')
ax.set_ylabel('Y axis')
ax.set_zlabel('Z axis')
plt.tight_layout()
plt.show()
```

Output:

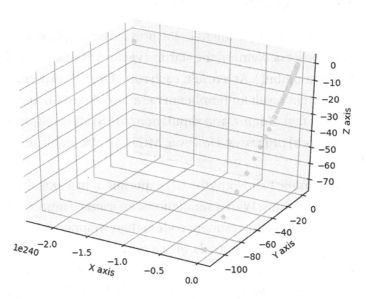

Figure 4-7. *3-D rendition of local minimum*

The code begins by importing matplotlib, mpl_toolkits, random, numpy, and scipy libraries. Function step() moves a vector in a direction (based on the gradient), by a step size. Function sigmoid_gradient() is the f'(sigmoid) returned as a point in 3-D space. Function mod_vector() ensures that an erroneous vector generated by the simulation is handled properly. The main block begins by creating a randomly generated 3-D vector [x, y, z] as a starting point for the simulation. It continues by creating a tolerance (precision). A smaller tolerance results in a more accurate result. A subplot is created to hold a 3-D rendering of the local minimum (Figure 4-7). The GD simulation creates a set of 3-D vectors influenced by the sigmoid gradient until the gradient is very small. The size (magnitude) of the gradient is calculated by the ED metric. The local minimum, as expected is close to 0.

Stabilizing Euclidean Distance Minimization with Monte Carlo Simulation

The Euclidean distance experiment in the previous example is anchored by a stochastic process. Namely, the starting vector v is stochastically generated by randomint(). As a result, each run of the GD experiment generates a different result for number of iterations. From Chapter 2, we already know that Monte Carlo simulation (MCS) efficiently models stochastic (random) processes. However, MCS can also stabilize stochastic experiments.

The code example first wraps the GD experiment in a loop that runs n number of simulations. With n simulations, an average number of iterations is calculated. The resultant code is then wrapped in another loop that runs m trials. With m trials, an average gap between each average number of iterations, is calculated. Gap is calculated by subtracting the minimum from the maximum average iteration. The smaller the gap, the more stable (accurate) the result. To increase accuracy, increase simulations (n). The only limitation is computing power. That is, running 1,000 simulations takes a lot more computing power than 100. Stable (accurate) results allow comparison to alternative experiments.

```python
import random, numpy as np
from scipy.spatial import distance

def step(v, direction, step_size):
    return [v_i + step_size * direction_i
            for v_i, direction_i in zip(v, direction)]

def sigmoid_gradient(v):
    return [v_i * (1-v_i) for v_i in v]

def mod_vector(v):
    for i, v_i in enumerate(v):
        if v_i == float("inf") or v_i == float("-inf"):
            v[i] = random.randint(-1, 1)
    return v

if __name__ == "__main__":
    trials= 10
    sims = 10
    avg_its = []
    for _ in range(trials):
        its = []
        for _ in range(sims):
            v = [random.randint(-10, 10) for i in range(3)]
            tolerance = 0.0000001
            iterations = 0
            while True:
                gradient = sigmoid_gradient(v)
                next_v = step(v, gradient, -0.01)
                v = mod_vector(v)
```

```
            next_v = mod_vector(next_v)
            test_v = distance.euclidean(v, next_v)
            if test_v < tolerance:
                break
            v = next_v
            iterations += 1
        its.append(iterations)
    a = round(np.mean(its))
    avg_its.append(a)
gap = np.max(avg_its) - np.min(avg_its)
print (trials, 'trials with', sims, 'simulations each:')
print ('gap', gap)
print ('avg iterations', round(np.mean(avg_its)))
```

Output:

```
10 trials with 10 simulations each:
gap 243.0
avg iterations 1031.0

10 trials with 100 simulations each:
gap 97.0
avg iterations 1087.0

10 trials with 1000 simulations each:
gap 13.0
avg iterations 1089.0
```

Output is for 10, 100, and 1,000 simulations. By running 1,000 simulations ten times (trials), the gap is down to 13. So, confidence is high that the number of iterations required to minimize the function is close to 1,089. We can further stabilize by wrapping the code in another loop to decrease variation in gap and number of iterations. However, computer processing time becomes an issue. Leveraging MCS for this type of experiment makes a strong case for cloud computing. It may be tough to get your head around this application of MCS, but it is a very powerful tool for working with and solving data science problems.

Substituting a NumPy Method to Hasten Euclidean Distance Minimization

Since numpy arrays are faster than Python lists, it follows that using a numpy method would be more efficient for calculating Euclidean distance. The code example substitutes np.linalg.norm() for distance.euclidean() to calculate Euclidean distance for the GD experiment.

```python
import matplotlib.pyplot as plt
from mpl_toolkits.mplot3d import Axes3D
import random, numpy as np

def step(v, direction, step_size):
    return [v_i + step_size * direction_i
            for v_i, direction_i in zip(v, direction)]

def sigmoid_gradient(v):
    return [v_i * (1-v_i) for v_i in v]

def round_v(v):
    return np.round(v, decimals=3)

if __name__ == "__main__":
    v = [random.randint(-10, 10) for i in range(3)]
    tolerance = 0.0000001
    iterations = 1
    fig = plt.figure('norm')
    ax = fig.add_subplot(111, projection='3d')
    while True:
        gradient = sigmoid_gradient(v)
        next_v = step(v, gradient, -0.01)
        round_gradient = round_v(gradient)
        xs = round_gradient[0]
        ys = round_gradient[1]
```

```
    zs = round_gradient[2]
    ax.scatter(xs, ys, zs, c='lime', marker='o')
    norm_v = np.linalg.norm(v)
    norm_next_v = np.linalg.norm(next_v)
    test_v = norm_v - norm_next_v
    if test_v < tolerance:
        break
    v = next_v
    iterations += 1
print ('minimum:', test_v, '\niterations:', iterations)
ax.set_xlabel('X axis')
ax.set_ylabel('Y axis')
ax.set_zlabel('Z axis')
plt.show()
```

Output:

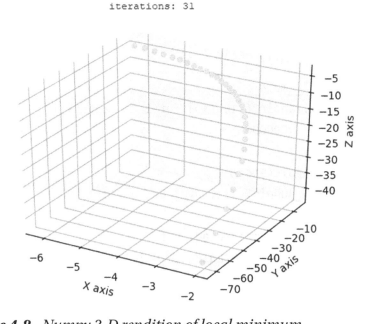

Figure 4-8. Numpy 3-D rendition of local minimum

The number of iterations is much lower at 31 (Figure 4-8). However, given that the GD experiment is stochastic, we can use MCS for objective comparison.

Using the same MCS methodology, the code example first wraps the GD experiment in a loop that runs n number of simulations. The resultant code is then wrapped in another loop that runs m trials.

```python
import random, numpy as np

def step(v, direction, step_size):
    return [v_i + step_size * direction_i
            for v_i, direction_i in zip(v, direction)]

def sigmoid_gradient(v):
    return [v_i * (1-v_i) for v_i in v]

def round_v(v):
    return np.round(v, decimals=3)

if __name__ == "__main__":
    trials= 10
    sims = 10
    avg_its = []
    for _ in range(trials):
        its = []
        for _ in range(sims):
            v = [random.randint(-10, 10) for i in range(3)]
            tolerance = 0.0000001
            iterations = 0
            while True:
                gradient = sigmoid_gradient(v)
                next_v = step(v, gradient, -0.01)
                norm_v = np.linalg.norm(v)
                norm_next_v = np.linalg.norm(next_v)
```

```
                test_v = norm_v - norm_next_v
                if test_v < tolerance:
                    break
                v = next_v
                iterations += 1
            its.append(iterations)
        a = round(np.mean(its))
        avg_its.append(a)
    gap = np.max(avg_its) - np.min(avg_its)
    print (trials, 'trials with', sims, 'simulations each:')
    print ('gap', gap)
    print ('avg iterations', round(np.mean(avg_its)))
```

Output:

```
10 trials with 10 simulations each:
gap 235.0
avg iterations 164.0

10 trials with 100 simulations each:
gap 141.0
avg iterations 200.0

10 trials with 1000 simulations each:
gap 27.0
avg iterations 193.0
```

Processing is much faster using numpy. The average number of iterations is close to 193. As such, using the numpy alternative for calculating Euclidean distance is more than five times faster!

Stochastic Gradient Descent Minimization and Maximization

Up to this point in the chapter, optimization experiments used batch GD. Batch GD computes the gradient using the whole dataset. Stochastic GD computes the gradient using a single sample, so it is computationally much faster. It is called stochastic GD because the gradient is randomly determined. However, unlike batch GD, stochastic GD is an approximation. If the exact gradient is required, stochastic GD is not optimal. Another issue with stochastic GD is that it can hover around the minimum forever without actually converging. So, it is important to plot progress of the simulation to see what is happening.

Let's change direction and optimize another important function–residual sum of squares (RSS). A RSS function is a statistical technique that measures the amount of error (variance) remaining between the regression function and the data set. Regression analysis is an algorithm that estimates relationships between variables. It is widely used for prediction and forecasting. It is also a popular modeling and predictive algorithm for data science applications.

The 1st code example generates a sample, runs the GD experiment n times, and processes the sample randomly:

```python
import matplotlib.pyplot as plt
import random, numpy as np

def rnd():
    return [random.randint(-10,10) for i in range(3)]

def random_vectors(n):
    ls = []
    for v in range(n):
        ls.append(rnd())
    return ls

def sos(v):
    return sum(v_i ** 2 for v_i in v)

def sos_gradient(v):
    return [2 * v_i for v_i in v]
```

```python
def in_random_order(data):
    indexes = [i for i, _ in enumerate(data)]
    random.shuffle(indexes)
    for i in indexes:
        yield data[i]

if __name__ == "__main__":
    v, x, y = rnd(), random_vectors(3), random_vectors(3)
    data = list(zip(x, y))
    theta = v
    alpha, value = 0.01, 0
    min_theta, min_value = None, float("inf")
    iterations_with_no_improvement = 0
    n, x = 30, 1
    for i, _ in enumerate(range(n)):
        y = np.linalg.norm(theta)
        plt.scatter(x, y, c='r')
        x = x + 1
        s = []
        for x_i, y_i in data:
            s.extend([sos(theta), sos(x_i), sos(y_i)])
        value = sum(s)
        if value < min_value:
            min_theta, min_value = theta, value
            iterations_with_no_improvement = 0
            alpha = 0.01
        else:
            iterations_with_no_improvement += 1
            alpha *= 0.9
        g = []
```

```
    for x_i, y_i in in_random_order(data):
        g.extend([sos_gradient(theta), sos_gradient(x_i),
                sos_gradient(y_i)])
        for v in g:
            theta = np.around(np.subtract(theta,alpha*np.
            array(v)),3)
        g = []
print ('minimum:', np.around(min_theta, 4),
      'with', i+1, 'iterations')
print ('iterations with no improvement:',
      iterations_with_no_improvement)
print ('magnitude of min vector:', np.linalg.norm(min_theta))
plt.show()
```

Output:

```
minimum: [ 0.609 -2.07   1.892] with 30 iterations
iterations with no improvement: 9
magnitude of min vector: 2.86974650448
```

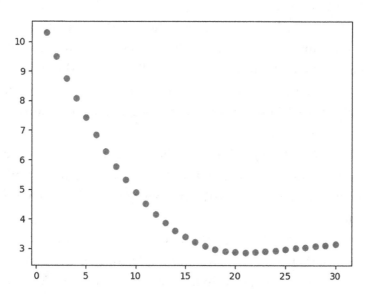

Figure 4-9. *RSS minimization*

The code begins by importing matplotlib, random, and numpy. It continues with function rnd(), which returns a list of random integers from –10 to 10. Function random_vectors() generates a list (random sample) of n numbers. Function sos() returns the RSS for a vector. Function sos_ gradient() returns the derivative (gradient) of RSS for a vector. Function in_random_order() generates a list of randomly shuffled indexes. This function adds the stochastic flavor to the GD algorithm. The main block begins by generating a random vector v as the starting point for the simulation. It continues by creating a sample of x and y vectors of size 3. Next, the vector is assigned to theta, which is a common name for a vector of some general probability distribution. We can call the vector anything we want, but a common data science problem is to find the value(s) of theta. The code continues with a fixed step size alpha, minimum theta value, minimum ending value, iterations with no improvement, number of simulations n, and a plot value for the x-coordinate (Figure 4-9).

The simulation begins by assigning y the magnitude of theta. Next, it plots the current x and y coordinates. The x-coordinate is incremented by 1 to plot the convergence to the minimum for each y-coordinate. The next block of code finds the RSS for each theta, and the sample of x and y values. This value determines if the simulation is hovering around the local minimum rather than converging. The final part of the code traverses the sample data points in random (stochastic) order, finds the gradient of theta, x and y, places these three values in list g, and traverses this vector to find the next theta value.

Whew! This is not simple, but this is how stochastic GD operates. Notice that the minimum generated is 2.87, which is not the true minimum of 0. So, stochastic GD requires few iterations but does not produce the true minimum.

The previous simulation can be refined by adjusting the algorithm for finding the next theta. In the previous example, the next theta is calculated for the gradient based on the current theta, x value, and y value for each sample. However, the actual new theta is based on the 3rd data point in the sample. So, the 2nd example is refined by taking the minimum theta from the entire sample rather than the 3rd data point:

```python
import matplotlib.pyplot as plt
import random, numpy as np

def rnd():
    return [random.randint(-10,10) for i in range(3)]

def random_vectors(n):
    ls = []
    for v in range(n):
        ls.append([random.randint(-10,10) for i in range(3)])
    return ls

def sos(v):
    return sum(v_i ** 2 for v_i in v)

def sos_gradient(v):
    return [2 * v_i for v_i in v]

def in_random_order(data):
    indexes = [i for i, _ in enumerate(data)]
    random.shuffle(indexes)
    for i in indexes:
        yield data[i]

if __name__ == "__main__":
    v, x, y = rnd(), random_vectors(3), random_vectors(3)
    data = list(zip(x, y))
    theta = v
    alpha, value = 0.01, 0
    min_theta, min_value = None, float("inf")
    iterations_with_no_improvement = 0
    n, x = 60, 1
```

```
for i, _ in enumerate(range(n)):
    y = np.linalg.norm(theta)
    plt.scatter(x, y, c='r')
    x = x + 1
    s = []
    for x_i, y_i in data:
        s.extend([sos(theta), sos(x_i), sos(y_i)])
    value = sum(s)
    if value < min_value:
        min_theta, min_value = theta, value
        iterations_with_no_improvement = 0
        alpha = 0.01
    else:
        iterations_with_no_improvement += 1
        alpha *= 0.9
    g, t, m = [], [], []
    for x_i, y_i in in_random_order(data):
        g.extend([sos_gradient(theta), sos_gradient(x_i),
                  sos_gradient(y_i)])
        m = np.around([np.linalg.norm(x) for x in g], 2)
        for v in g:
            theta = np.around(np.subtract(theta,alpha*np.
            array(v)),3)
            t.append(np.around(theta,2))
        mm = np.argmin(m)
        theta = t[mm]
        g, m, t = [], [], []
print ('minimum:', np.around(min_theta, 4),
       'with', i+1, 'iterations')
```

```
print ('iterations with no improvement:',
        iterations_with_no_improvement)
print ('magnitude of min vector:', np.linalg.norm(min_theta))
plt.show()
```

Output:

```
minimum: [ 0.26   0.26   0.26] with 60 iterations
iterations with no improvement: 3
magnitude of min vector: 0.450333209968
```

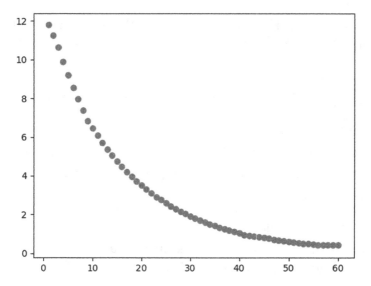

Figure 4-10. *Modified RSS minimization*

The only difference in the code is toward the bottom where the minimum theta is calculated (Figure 4-10). Although it took 60 iterations, the minimum is much closer to 0 and much more stable. That is, the prior example deviates quite a bit more each time the experiment is run.

The 3rd example finds the maximum:

```python
import matplotlib.pyplot as plt
import random, numpy as np

def rnd():
    return [random.randint(-10,10) for i in range(3)]

def random_vectors(n):
    ls = []
    for v in range(n):
        ls.append([random.randint(-10,10) for i in range(3)])
    return ls

def sos_gradient(v):
    return [2 * v_i for v_i in v]

def negate(function):
    def new_function(*args, **kwargs):
        return np.negative(function(*args, **kwargs))
    return new_function

def in_random_order(data):
    indexes = [i for i, _ in enumerate(data)]
    random.shuffle(indexes)
    for i in indexes:
        yield data[i]

if __name__ == "__main__":
    v, x, y = rnd(), random_vectors(3), random_vectors(3)
    data = list(zip(x, y))
    theta, alpha = v, 0.01
    neg_gradient = negate(sos_gradient)
    n, x = 100, 1
```

```
for i, row in enumerate(range(n)):
    y = np.linalg.norm(theta)
    plt.scatter(x, y, c='r')
    x = x + 1
    g = []
    for x_i, y_i in in_random_order(data):
        g.extend([neg_gradient(theta), neg_gradient(x_i),
                neg_gradient(y_i)])
        for v in g:
            theta = np.around(np.subtract(theta,alpha*np.
            array(v)),3)
        g = []
print ('maximum:', np.around(theta, 4),
        'with', i+1, 'iterations')
print ('magnitude of max vector:', np.linalg.norm(theta))
plt.show()
```

Output:

```
maximum: [-1521.6    4178.212 -3038.379] with 100 iterations
magnitude of max vector: 5385.57972967
```

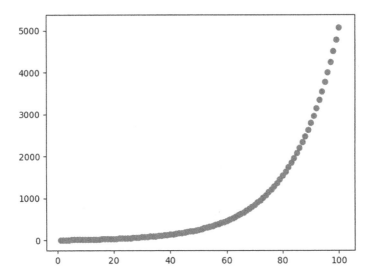

Figure 4-11. *RSS maximization*

The only difference in the code from the 1st example is the negate()
function, which negates the gradient to find the maximum. Since the
maximum of RSS is infinity (we can tell by the visualization in Figure 4-11),
we can stop at 100 iterations. Try 1,000 iterations and see what happens.

CHAPTER 5

Working with Data

Working with data details the earliest processes of data science problem solving. The 1st step is to identify the problem, which determines all else that needs to be done. The 2nd step is to gather data. The 3rd step is to wrangle (munge) data, which is critical. Wrangling is getting data into a form that is useful for machine learning and other data science problems. Of course, wrangled data will probably have to be cleaned. The 4th step is to visualize the data. Visualization helps you get to know the data and, hopefully, identify patterns.

One-Dimensional Data Example

The code example generates visualizations of two very common data distributions–uniform and normal. The uniform distribution has constant probability. That is, all events that belong to the distribution are equally probable. The normal distribution is symmetrical about the center, which means that 50% of its values are less than the mean and 50% of its values are greater than the mean. Its shape resembles a bell curve. The normal distribution is extremely important because it models many naturally occurring events.

© David Paper 2018
D. Paper, *Data Science Fundamentals for Python and MongoDB*,
https://doi.org/10.1007/978-1-4842-3597-3_5

```python
import matplotlib.pyplot as plt
import numpy as np

if __name__ == "__main__":
    plt.figure('Uniform Distribution')
    uniform = np.random.uniform(-3, 3, 1000)
    count, bins, ignored = plt.hist(uniform, 20, facecolor='lime')
    plt.xlabel('Interval: [-3, 3]')
    plt.ylabel('Frequency')
    plt.title('Uniform Distribution')
    plt.axis([-3,3,0,100])
    plt.grid(True)
    plt.figure('Normal Distribution')
    normal = np.random.normal(0, 1, 1000)
    count, bins, ignored = plt.hist(normal, 20,
    facecolor='fuchsia')
    plt.xlabel('Interval: [-3, 3]')
    plt.ylabel('Frequency')
    plt.title('Normal Distribution')
    plt.axis([-3,3,0,140])
    plt.grid(True)
    plt.show()
```

Output:

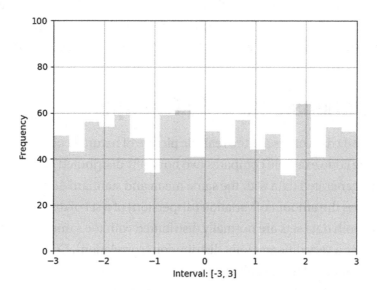

Figure 5-1. *Uniform distribution*

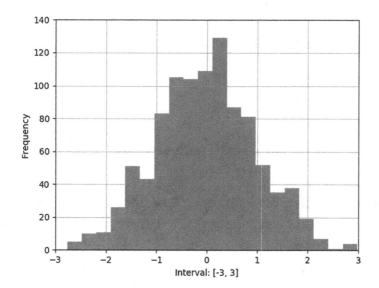

Figure 5-2. *Normal distribution*

The code example begins by importing matplotlib and numpy. The main block begins by creating a figure and data for a uniform distribution. Next, a histogram is created and plotted based on the data. A figure for a normal distribution is then created and plotted. See Figures 5-1 and 5-2.

Two-Dimensional Data Example

Modeling 2-D data offers a more realistic picture of naturally occurring events. The code example compares two normally distributed distributions of randomly generated data with the same mean and standard deviation (SD). SD measures the amount of variation (dispersion) of a set of data values. Although both data sets are normally distributed with the same mean and SD, each has a very different joint distribution (correlation). Correlation is the interdependence of two variables.

```
import matplotlib.pyplot as plt
import matplotlib.gridspec as gridspec
import numpy as np, random
from scipy.special import ndtri

def inverse_normal_cdf(r):
    return ndtri(r)

def random_normal():
    return inverse_normal_cdf(random.random())

def scatter(loc):
    plt.scatter(xs, ys1, marker='.', color='black', label='ys1')
    plt.scatter(xs, ys2, marker='.', color='gray',  label='ys2')
    plt.xlabel('xs')
    plt.ylabel('ys')
    plt.legend(loc=loc)
    plt.tight_layout()
```

```python
if __name__ == "__main__":
    xs = [random_normal() for _ in range(1000)]
    ys1 = [ x + random_normal() / 2 for x in xs]
    ys2 = [-x + random_normal() / 2 for x in xs]
    gs = gridspec.GridSpec(2, 2)
    fig = plt.figure()
    ax1 = fig.add_subplot(gs[0,0])
    plt.title('ys1 data')
    n, bins, ignored = plt.hist(ys1, 50, normed=1,
                                facecolor='chartreuse',
                                alpha=0.75)
    ax2 = fig.add_subplot(gs[0,1])
    plt.title('ys2 data')
    n, bins, ignored = plt.hist(ys2, 50, normed=1,
                                facecolor='fuchsia',
                                alpha=0.75)
    ax3 = fig.add_subplot(gs[1,:])
    plt.title('Correlation')
    scatter(6)
    print (np.corrcoef(xs, ys1)[0, 1])
    print (np.corrcoef(xs, ys2)[0, 1])
    plt.show()
```

Output:

0.907683439554
-0.896109957488

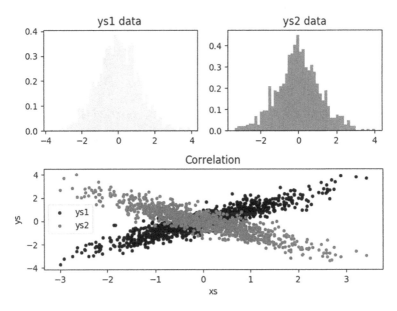

Figure 5-3. *Subplot of normal distributions and correlation*

The code example begins by importing matplotlib, numpy, random, and scipy libraries. Method gridspec specifies the geometry of a grid where a subplot will be placed. Method ndtri returns the standard normal cumulative distribution function (CDF). CDF is the probability that a random variable X takes on a value less than or equal to x, where x represents the area under a normal distribution. The code continues with three functions. Function inverse_normal_cdf() returns the CDF based on a random variable. Function random_normal() calls function inverse_normal_cdf() with a randomly generated value X and returns the CDF. Function scatter() creates a scatter plot. The main block begins by

134

creating randomly generated x and y values xs, ys1, and ys2. A gridspec() is created to hold the distributions. Histograms are created for xs, ys1 and xs, ys2 data, respectively. Next, a correlation plot is created for both distributions. Finally, correlations are generated for the two distributions. Figure 5-3 shows plots.

The code example spawns two important lessons. First, creating a set of randomly generated numbers with ndtri() creates a normally distributed dataset. That is, function ndtri() returns the CDF of a randomly generated value. Second, two normally distributed datasets are not necessarily similar even though they look alike. In this case, the correlations are opposite. So, visualization and correlations are required to demonstrate the difference between the datasets.

Data Correlation and Basic Statistics

Correlation is the extent that two or more variables fluctuate (move) together. A correlation matrix is a table showing correlation coefficients between sets of variables. Correlation coefficients measure strength of association between two or more variables.

The code example creates three datasets with x and y coordinates, calculates correlations, and plots. The 1st dataset represents a positive correlation; the 2nd, a negative correlation; and the 3rd, a weak correlation.

```
import random, numpy as np
import matplotlib.pyplot as plt
import matplotlib.gridspec as gridspec

if __name__ == "__main__":
    np.random.seed(0)
    x = np.random.randint(0, 50, 1000)
    y = x + np.random.normal(0, 10, 1000)
```

```
print ('highly positive:\n', np.corrcoef(x, y))
gs = gridspec.GridSpec(2, 2)
fig = plt.figure()
ax1 = fig.add_subplot(gs[0,0])
plt.title('positive correlation')
plt.scatter(x, y, color='springgreen')
y = 100 - x + np.random.normal(0, 10, 1000)
print ('\nhighly negative:\n', np.corrcoef(x, y))
ax2 = fig.add_subplot(gs[0,1])
plt.title('negative correlation')
plt.scatter(x, y, color='crimson')
y = np.random.normal(0, 10, 1000)
print ('\nno/weak:\n', np.corrcoef(x, y))
ax3 = fig.add_subplot(gs[1,:])
plt.title('weak correlation')
plt.scatter(x, y, color='peachpuff')
plt.tight_layout()
plt.show()
```

Output:

```
highly positive:
[[ 1.          0.82777267]
 [ 0.82777267  1.         ]]

highly negative:
[[ 1.         -0.8350955]
 [-0.8350955  1.        ]]

no/weak:
[[ 1.          0.00962676]
 [ 0.00962676  1.         ]]
```

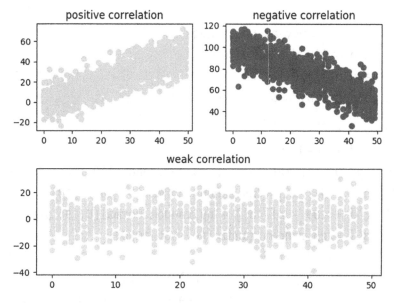

Figure 5-4. *Subplot of correlations*

The code example begins by importing random, numpy, and matplotlib libraries. The main block begins by generating x and y coordinates with a positive correlation and displaying the correlation matrix. It continues by creating a grid to hold the subplot, the 1st subplot grid, and a scatterplot. Next, x and y coordinates are created with a negative correlation and the correlation matrix is displayed. The 2nd subplot grid is created and plotted. Finally, x and y coordinates are created with a weak correlation and the correlation matrix is displayed. The 3rd subplot grid is created and plotted, and all three scatterplots are displayed. Figure 5-4 shows the plots.

Pandas Correlation and Heat Map Examples

Pandas is a Python package that provides fast, flexible, and expressive data structures to make working with virtually any type of data easy, intuitive, and practical in real-world data analysis. A DataFrame (df) is a 2-D labeled data structure and the most commonly used object in pandas.

The 1st code example creates a correlation matrix with an associated visualization:

```python
import random, numpy as np, pandas as pd
import matplotlib.pyplot as plt
import matplotlib.cm as cm
import matplotlib.colors as colors

if __name__ == "__main__":
    np.random.seed(0)
    df = pd.DataFrame({'a': np.random.randint(0, 50, 1000)})
    df['b'] = df['a'] + np.random.normal(0, 10, 1000)
    df['c'] = 100 - df['a'] + np.random.normal(0, 5, 1000)
    df['d'] = np.random.randint(0, 50, 1000)
    colormap = cm.viridis
    colorlist = [colors.rgb2hex(colormap(i))
                    for i in np.linspace(0, 1, len(df['a']))]
    df['colors'] = colorlist
    print (df.corr())
    pd.plotting.scatter_matrix(df, c=df['colors'],
                                    diagonal='d',
                                    figsize=(10, 6))

    plt.show()
```

Output:

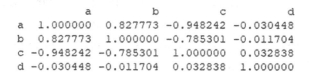

```
            a          b          c          d
a   1.000000   0.827773  -0.948242  -0.030448
b   0.827773   1.000000  -0.785301  -0.011704
c  -0.948242  -0.785301   1.000000   0.032838
d  -0.030448  -0.011704   0.032838   1.000000
```

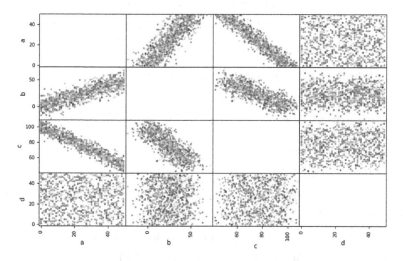

Figure 5-5. *Correlation matrix visualization*

The code example begins by importing random, numpy, pandas, and matplotlib libraries. The main block begins by creating a df with four columns populated by various random number possibilities. It continues by creating a color map of the correlations between each column, printing the correlation matrix, and plotting the color map (Figure 5-5).

We can see from the correlation matrix that the most highly correlated variables are a and b (0.83), a and c (-0.95), and b and c (-0.79). From the color map, we can see that a and b are positively correlated, a and c are negatively correlated, and b and c are negatively correlated. However, the actual correlation values are not apparent from the visualiztion.

A Heat map is a graphical representation of data where individual values in a matrix are represented as colors. It is a popular visualization technique in data science. With pandas, a Heat map provides a sophisticated visualization of correlations where each variable is represented by its own color.

The 2nd code example uses a Heat map to visualize variable correlations. You need to install library seaborn if you don't already have it installed on your computer (e.g., pip install seaborn).

```
import random, numpy as np, pandas as pd
import matplotlib.pyplot as plt
import seaborn as sns

if __name__ == "__main__":
    np.random.seed(0)
    df = pd.DataFrame({'a': np.random.randint(0, 50, 1000)})
    df['b'] = df['a'] + np.random.normal(0, 10, 1000)
    df['c'] = 100 - df['a'] + np.random.normal(0, 5, 1000)
    df['d'] = np.random.randint(0, 50, 1000)
    plt.figure()
    sns.heatmap(df.corr(), annot=True, cmap='OrRd')
    plt.show()
```

Output:

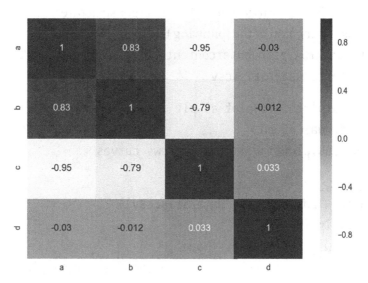

Figure 5-6. *Heat map*

The code begins by importing random, numpy, pandas, matplotlib, and seaborn libraries. Seaborn is a Python visualization library based on matplotlib. The main block begins by generating four columns of data (variables), and plots a Heat map (Figure 5-6). Attribute cmap uses a colormap. A list of matplotlib colormaps can be found at: `https://matplotlib.org/examples/color/colormaps_reference.html`.

Various Visualization Examples

The 1st code example introduces the Andrews curve, which is a way to visualize structure in high-dimensional data. Data for this example is the Iris dataset, which is one of the best known in the pattern recognition literature. The Iris dataset consists of three different types of irises' (Setosa, Versicolour, and Virginica) petal and sepal lengths.

Andrews curves allow multivariate data plotting as a large number of curves that are created using the attributes (variable) of samples as coefficients. By coloring the curves differently for each class, it is possible to visualize data clustering. Curves belonging to samples of the same class will usually be closer together and form larger structures. Raw data for the iris dataset is located at the following URL:

https://raw.githubusercontent.com/pandas-dev/pandas/master/pandas/tests/data/iris.csv

```
import matplotlib.pyplot as plt
import pandas as pd
from pandas.plotting import andrews_curves

if __name__ == "__main__":
    data = pd.read_csv('data/iris.csv')
    plt.figure()
    andrews_curves(data, 'Name',
                   color=['b','mediumspringgreen','r'])
    plt.show()
```

Output:

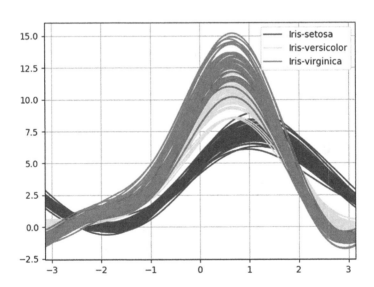

Figure 5-7. *Andrews curves*

The code example begins by importing matplotlib and pandas. The main block begins by reading the iris dataset into pandas df data. Next, Andrews curves are plotted for each class–Iris-setosa, Iris-versicolor, and Iris-virginica (Figure 5-7). From this visualization, it is difficult to see which attributes distinctly define each class.

The 2nd code example introduces parallel coordinates:

```
import matplotlib.pyplot as plt
import pandas as pd
from pandas.plotting import parallel_coordinates

if __name__ == "__main__":
    data = pd.read_csv('data/iris.csv')
    plt.figure()
    parallel_coordinates(data, 'Name',
                          color=['b','mediumspringgreen','r'])
    plt.show()
```

Output:

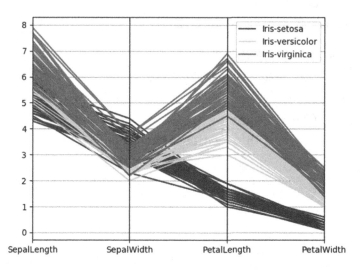

Figure 5-8. *Parallel coordinates*

Parallel coordinates is another technique for plotting multivariate data. It allows visualization of clusters in data and estimation of other statistics visually. Points are represented as connected line segments. Each vertical line represents one attribute. One set of connected line segments represents one data point. Points that tend to cluster appear closer together.

The code example begins by importing matplotlib and pandas. The main block begins by reading the iris dataset into pandas df data. Next, parallel coordinates are plotted for each class (Figure 5-8). From this visualization, attributes PetalLength and PetalWidth are most distinct for the three species (classes of Iris). So, PetalLength and PetalWidth are the best classifiers for species of Iris. Andrews curves just don't clearly provide this important information.

Here is a useful URL:

`http://wilkelab.org/classes/SDS348/2016_spring/worksheets/class9.html`

The 3rd code example introduces RadViz:

```
import matplotlib.pyplot as plt
import pandas as pd
from pandas.plotting import radviz

if __name__ == "__main__":
    data = pd.read_csv('data/iris.csv')
    plt.figure()
    radviz(data, 'Name',
           color=['b','mediumspringgreen','r'])
    plt.show()
```

Output:

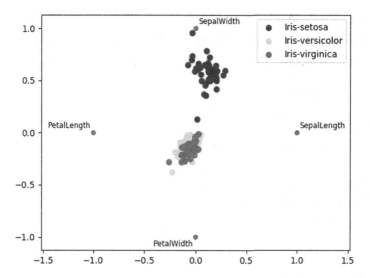

Figure 5-9. _RadVis_

RadVis is yet another technique for visualizing multivariate data. The code example begins by importing matplotlib and pandas. The main block begins by reading the iris dataset into pandas df data. Next, RadVis coordinates are plotted for each class (Figure 5-9). With this visualization, it is not easy to see any distinctions. So, the parallel coordinates technique appears to be the best of the three in terms of recognizing variation (for this example).

Cleaning a CSV File with Pandas and JSON

The code example loads a dirty CSV file into a Pandas df and displays to locate bad data. It then loads the same CSV file into a list of dictionary elements for cleaning. Finally, the cleansed data is saved to JSON.

```python
import csv, pandas as pd, json

def to_dict(d):
    return [dict(row) for row in d]

def dump_json(f, d):
    with open(f, 'w') as f:
        json.dump(d, f)

def read_json(f):
    with open(f) as f:
        return json.load(f)

if __name__ == "__main__":
    df = pd.read_csv("data/audio.csv")
    print (df, '\n')
    data = csv.DictReader(open('data/audio.csv'))
    d = to_dict(data)
    for row in d:
        if (row['pno'][0] not in ['a', 'c', 'p', 's']):
            if (row['pno'][0] == '8'):
                row['pno'] = 'a' + row['pno']
            elif (row['pno'][0] == '7'):
                row['pno'] = 'p' + row['pno']
            elif (row['pno'][0] == '5'):
                row['pno'] = 's' + row['pno']
```

```python
        if (row['color']) == '-':
            row['color'] = 'silver'
        if row['model'] == '-':
            row['model'] = 'S1'
        if (row['mfg']) == '100':
            row['mfg'] = 'Linn'
        if (row['desc'] == '0') and row['pno'][0] == 'p':
            row['desc'] = 'preamplifier'
        elif (row['desc'] == '-') and row['pno'][0] == 's':
            row['desc'] = 'speakers'
        if (row['price'][0] == '$'):
            row['price'] =\
            row['price'].translate({ord(i): None for i in '$,.'})
json_file = 'data/audio.json'
dump_json(json_file, d)
data = read_json(json_file)
for i, row in enumerate(data):
    if i < 5:
        print (row)
```

Output:

```
     pno       color          mfg       model          desc         price
0   a89632     silver   Jeff Roland     JR 302      amplifier       10000
1   a85412     silver     AVM Audio     MA 3.2      amplifier        5890
2    87425      black       Gryphon    Antileon     amplifier       13999
3   a85879  champagne     Parasound       JC 1      amplifier        4850
4    82415       gray      SimAudio       400M      amplifier    $4,500.00
5   a81111      black         Krell    KSA 300s     amplifier        4250
6   c10001     silver           100        CD12           cdp       20000
7   c11023     silver         Hegel     Mohican           cdp      $5,000
8   p70022          -     AVM Audio      PA 3.2  preamplifier        3998
9    79999      black     Sovereign    Director  preamplifier        8250
10  78787      silver      SimAudio         P-8  preamplifier        7300
11  p77777          -          Linn        KK 1  preamplifier        8950
12  p71010       gray         Krell         KSL             0        1499
13  70027       black        Classe     SSP-600  preamplifier        5999
14  p71000     silver       Boulder      B 1012  preamplifier       11700
15  55555      cherry         Thiel     CS 2.45E      speakers        7999
16  s51212      cherry       Harbeth      H 40.1      speakers   $12,995.00
17  s50000   composite       Magico           -      speakers       35000
18  59999        gray        Wilson    Sasha W/P      speakers       21500
19  s53232      silver  YG Acoustics   Anat III       speakers       40000

{'pno': 'a89632', 'color': 'silver', 'mfg': 'Jeff Roland', 'model': 'JR 302', 'desc': 'amplifier', 'price': '10000'}
{'pno': 'a85412', 'color': 'silver', 'mfg': 'AVM Audio', 'model': 'MA 3.2', 'desc': 'amplifier', 'price': '5890'}
{'pno': 'a87425', 'color': 'black', 'mfg': 'Gryphon', 'model': 'Antileon', 'desc': 'amplifier', 'price': '13999'}
{'pno': 'a85879', 'color': 'champagne', 'mfg': 'Parasound', 'model': 'JC 1', 'desc': 'amplifier', 'price': '4850'}
{'pno': 'a82415', 'color': 'gray', 'mfg': 'SimAudio ', 'model': '400M', 'desc': 'amplifier', 'price': '450000 '}
```

The code example begins by importing csv, pandas, and json libraries. Function to_dict() converts a list of OrderedDict elements to a list of regular dictionary elements for easier processing. Function dump_json() saves data to a JSON file. Function read_json() reads JSON data into a Python list. The main block begins by loading a CSV file into a Pandas df and displaying it to visualize dirty data. It continues by loading the same CSV file into a list of dictionary elements for easier cleansing. Next, all dirty data is cleansed. The code continues by saving the cleansed data to JSON file audio.json. Finally, audio.json is loaded and a few records are displayed to ensure that everything worked properly.

Slicing and Dicing

Slicing and dicing is breaking data into smaller parts or views to better understand and present it as information in a variety of different and useful ways. A slice in multidimensional arrays is a column of data corresponding to a single value for one or more members of the dimension of interest. While a slice filters on a particular attribute, a dice is like a zoom feature that selects a subset of all dimensions, but only for specific values of the dimension.

The code example loads audio.json into a Pandas df, slices data by column and row, and displays:

```
import pandas as pd

if __name__ == "__main__":
    df = pd.read_json("data/audio.json")
    amps = df[df.desc == 'amplifier']
    print (amps, '\n')
    price = df.query('price >= 40000')
    print (price, '\n')
    between = df.query('4999 < price < 6000')
    print (between, '\n')
```

```
row = df.loc[[0, 10, 19]]
print (row)
```

Output:

```
        color           desc          mfg     model      pno    price
0       silver     amplifier   Jeff Roland    JR 302   a89632    10000
1       silver     amplifier     AVM Audio    MA 3.2   a85412     5890
2        black     amplifier       Gryphon  Antileon   a87425    13999
3    champagne     amplifier     Parasound      JC 1   a85879     4850
4         gray     amplifier      SimAudio      400M   a82415   450000
5        black     amplifier         Krell  KSA 300s   a81111     4250

        color           desc          mfg     model      pno    price
4         gray     amplifier      SimAudio      400M   a82415   450000
16      cherry      speakers       Harbeth    H 40.1   s51212  1299500
19      silver      speakers  YG Acoustics  Anat III   s53232    40000

        color           desc          mfg     model      pno  price
1       silver     amplifier     AVM Audio    MA 3.2   a85412   5890
7       silver           cdp         Hegel   Mohican   c11023   5000
13       black  preamplifier        Classe   SSP-600   p70027   5999

        color           desc          mfg     model      pno  price
0       silver     amplifier   Jeff Roland    JR 302   a89632  10000
10      silver  preamplifier      SimAudio       P-8   p78787   7300
19      silver      speakers  YG Acoustics  Anat III   s53232  40000
```

The code example begins by importing Pandas. The main block begins by loading audio.json into a Pandas df. Next, the df is sliced by amplifier from the desc column. The code continues by slicing by the price column for equipment more expensive than $40,000. The next slice is by price column for equipment between $5,000 and $6,000. The final slice is by rows 0, 10, and 19.

Data Cubes

A data cube is an n-dimensional array of values. Since it is hard to conceptualize an n-dimensional cube, most are 3-D in practice.

Let's build a cube that holds three stocks–GOOGL, AMZ, and MKL. For each stock, include five days of data. Each day includes data for open, high, low, close, adj close, and volume values. So, the three dimensions are stock, day, and values. Data was garnered from actual stock quotes.

149

The code example creates a cube, saves it to a JSON file, reads the JSON, and displays some information:

```python
import json

def dump_json(f, d):
    with open(f, 'w') as f:
        json.dump(d, f)

def read_json(f):
    with open(f) as f:
        return json.load(f)

def rnd(n):
    return '{:.2f}'.format(n)

if __name__ == "__main__":
    d = dict()
    googl = dict()
    googl['2017-09-25'] =\
    {'Open':939.450012, 'High':939.750000, 'Low':924.510010,
     'Close':934.280029, 'Adj Close':934.280029,
     'Volume':1873400}
    googl['2017-09-26'] =\
    {'Open':936.690002, 'High':944.080017, 'Low':935.119995,
     'Close':937.429993, 'Adj Close':937.429993,
     'Volume':1672700}
    googl['2017-09-27'] =\
    {'Open':942.739990, 'High':965.429993, 'Low':941.950012,
     'Close':959.900024, 'Adj Close':959.900024,
     'Volume':2334600}
    googl['2017-09-28'] =\
    {'Open':956.250000, 'High':966.179993, 'Low':955.549988,
     'Close':964.809998, 'Adj Close':964.809998, 'Volume':1400900}
```

```
googl['2017-09-29'] =\
{'Open':966.000000, 'High':975.809998, 'Low':966.000000,
 'Close':973.719971, 'Adj Close':973.719971,
 'Volume':2031100}
amzn = dict()
amzn['2017-09-25'] =\
{'Open':949.309998, 'High':949.419983, 'Low':932.890015,
 'Close':939.789978, 'Adj Close':939.789978,
 'Volume':5124000}
amzn['2017-09-26'] =\
{'Open':945.489990, 'High':948.630005, 'Low':931.750000,
 'Close':937.429993, 'Adj Close':938.599976,
 'Volume':3564800}
amzn['2017-09-27'] =\
{'Open':948.000000, 'High':955.299988, 'Low':943.299988,
 'Close':950.869995, 'Adj Close':950.869995,
 'Volume':3148900}
amzn['2017-09-28'] =\
{'Open':951.859985, 'High':959.700012, 'Low':950.099976,
 'Close':956.400024, 'Adj Close':956.400024,
 'Volume':2522600}
amzn['2017-09-29'] =\
{'Open':960.109985, 'High':964.830017, 'Low':958.380005,
 'Close':961.349976, 'Adj Close':961.349976,
 'Volume':2543800}
mkl = dict()
mkl['2017-09-25'] =\
{'Open':1056.199951, 'High':1060.089966, 'Low':1047.930054,
 'Close':1050.250000, 'Adj Close':1050.250000,
 'Volume':23300}
```

```
mkl['2017-09-26'] =\
{'Open':1052.729980, 'High':1058.520020, 'Low':1045.000000,
 'Close':1045.130005, 'Adj Close':1045.130005,
 'Volume':25800}
mkl['2017-09-27'] =\
{'Open':1047.560059, 'High':1069.099976, 'Low':1047.010010,
 'Close':1064.040039, 'Adj Close':1064.040039,
 'Volume':21100}
mkl['2017-09-28'] =\
{'Open':1064.130005, 'High':1073.000000, 'Low':1058.079956,
 'Close':1070.550049, 'Adj Close':1070.550049,
 'Volume':23500}
mkl['2017-09-29'] =\
{'Open':1068.439941, 'High':1073.000000, 'Low':1060.069946,
 'Close':1067.979980, 'Adj Close':1067.979980 ,
 'Volume':20700}
d['GOOGL'], d['AMZN'], d['MKL'] = googl, amzn, mkl
json_file = 'data/cube.json'
dump_json(json_file, d)
d = read_json(json_file)
s = ' '    .
print ('\'Adj Close\' slice:')
print (10*s, 'AMZN', s, 'GOOGL', s, 'MKL')
print ('Date')
print ('2017-09-25', rnd(d['AMZN']['2017-09-25']
['Adj Close']),
       rnd(d['GOOGL']['2017-09-25']['Adj Close']),
       rnd(d['MKL']['2017-09-25']['Adj Close']))
print ('2017-09-26', rnd(d['AMZN']['2017-09-26']
['Adj Close']),
```

```
        rnd(d['GOOGL']['2017-09-26']['Adj Close']),
        rnd(d['MKL']['2017-09-26']['Adj Close']))
  print ('2017-09-27', rnd(d['AMZN']['2017-09-27']
        ['Adj Close']),
        rnd(d['GOOGL']['2017-09-27']['Adj Close']),
        rnd(d['MKL']['2017-09-27']['Adj Close']))
  print ('2017-09-28', rnd(d['AMZN']['2017-09-28']
        ['Adj Close']),
        rnd(d['GOOGL']['2017-09-28']['Adj Close']),
        rnd(d['MKL']['2017-09-28']['Adj Close']))
  print ('2017-09-29', rnd(d['AMZN']['2017-09-29']
        ['Adj Close']),
        rnd(d['GOOGL']['2017-09-29']['Adj Close']),
        rnd(d['MKL']['2017-09-29']['Adj Close']))
```

Output:

```
'Adj Close' slice:
              AMZN    GOOGL    MKL
Date
2017-09-25 939.79 934.28 1050.25
2017-09-26 938.60 937.43 1045.13
2017-09-27 950.87 959.90 1064.04
2017-09-28 956.40 964.81 1070.55
2017-09-29 961.35 973.72 1067.98
```

The code example begins by importing json. Function dump_json() and read_json() save and read JSON data respectively. The main block creates a cube by creating a dictionary d, dictionaries for each stock, and adding data by day and attribute to each stock dictionary. The code continues by saving the cube to JSON file cube.json. Finally, the code reads cube.json and displays a slice from the cube.

Data Scaling and Wrangling

Data scaling is changing type, spread, and/or position to compare data that are otherwise incomparable. Data scaling is very common in data science. Mean centering is the 1st technique, which transforms data by subtracting out the mean. Normalization is the 2nd technique, which transforms data to fall within the range between 0 and 1. Standardization is the 3rd technique, which transforms data to zero mean and unit variance (SD = 1), which is commonly referred to as standard normal.

The 1st code example generates and centers a normal distribution:

```python
import numpy as np
import matplotlib.pyplot as plt

def rnd_nrml(m, s, n):
    return np.random.normal(m, s, n)

def ctr(d):
    return [x-np.mean(d) for x in d]

if __name__ == "__main__":
    mu, sigma, n, c1, c2, b = 10, 15, 100, 'pink',\
                              'springgreen', True
    s = rnd_nrml(mu, sigma, n)
    plt.figure()
    ax = plt.subplot(211)
    ax.set_title('normal distribution')
    count, bins, ignored = plt.hist(s, 30, color=c1, normed=b)
    sc = ctr(s)
    ax = plt.subplot(212)
    ax.set_title('normal distribution "centered"')
    count, bins, ignored = plt.hist(sc, 30, color=c2, normed=b)
    plt.tight_layout()
    plt.show()
```

Output:

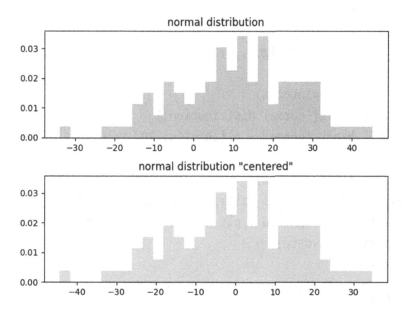

Figure 5-10. *Subplot for centering data*

The code example begins by importing numpy and matplotlib.
Function rnd_nrml() generates a normal distribution based on mean
(mu), SD (sigma), and n number of data points. Function ctr() subtracts
out the mean from every data point. The main block begins by creating
the normal distribution. The code continues by plotting the original and
centered distributions (Figure 5-10). Notice that the distributions are
exactly the same, but the 2nd distribution is centered with mean of 0.

The 2nd code example generates and normalizes a normal distribution:

```
import numpy as np
import matplotlib.pyplot as plt

def rnd_nrml(m, s, n):
    return np.random.normal(m, s, n)

def nrml(d):
    return [(x-np.amin(d))/(np.amax(d)-np.amin(d)) for x in d]
```

```
if __name__ == "__main__":
    mu, sigma, n, c1, c2, b = 10, 15, 100, 'orchid',\
                                'royalblue', True
    s = rnd_nrml(mu, sigma, n)
    plt.figure()
    ax = plt.subplot(211)
    ax.set_title('normal distribution')
    count, bins, ignored = plt.hist(s, 30, color=c1, normed=b)
    sn = nrml(s)
    ax = plt.subplot(212)
    ax.set_title('normal distribution "normalized"')
    count, bins, ignored = plt.hist(sn, 30, color=c2, normed=b)
    plt.tight_layout()
    plt.show()
```

Output:

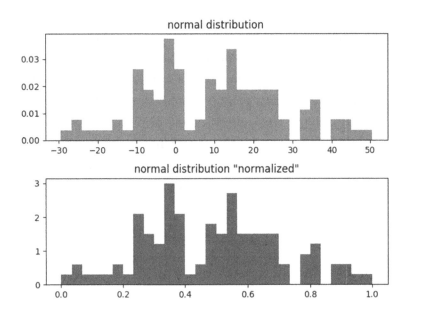

Figure 5-11. *Subplot for normalizing data*

The code example begins by importing numpy and matplotlib. Function rnd_nrml() generates a normal distribution based on mean (mu), SD (sigma), and n number of data points. Function nrml() transforms data to fall within the range between 0 and 1. The main block begins by creating the normal distribution. The code continues by plotting the original and normalized distributions (Figure 5-11). Notice that the distributions are exactly the same, but the 2nd distribution is normalized between 0 and 1.

The 3rd code example transforms data to zero mean and unit variance (standard normal):

```python
import numpy as np, csv
import matplotlib.pyplot as plt

def rnd_nrml(m, s, n):
    return np.random.normal(m, s, n)

def std_nrml(d, m, s):
    return [(x-m)/s for x in d]

if __name__ == "__main__":
    mu, sigma, n, b = 0, 1, 1000, True
    c1, c2 = 'peachpuff', 'lime'
    s = rnd_nrml(mu, sigma, n)
    plt.figure(1)
    plt.title('standard normal distribution')
    count, bins, ignored = plt.hist(s, 30, color=c1, normed=b)
    plt.plot(bins, 1/(sigma * np.sqrt(2 * np.pi)) *
            np.exp( - (bins - mu)**2 / (2 * sigma**2) ),
            linewidth=2, color=c2)
    start1, start2 = 5, 600
    mu1, sigma1, n, b = 10, 15, 500, True
    x1 = np.arange(start1, n+start1, 1)
    y1 = rnd_nrml(mu1, sigma1, n)
    mu2, sigma2, n, b = 25, 5, 500, True
```

```
x2 = np.arange(start2, n+start2, 1)
y2 = rnd_nrml(mu2, sigma2, n)
plt.figure(2)
ax = plt.subplot(211)
ax.set_title('dataset1 (mu=10, sigma=15)')
count, bins, ignored = plt.hist(y1, 30, color='r', normed=b)
ax = plt.subplot(212)
ax.set_title('dataset2 (mu=5, sigma=5)')
count, bins, ignored = plt.hist(y2, 30, color='g', normed=b)
plt.tight_layout()
plt.figure(3)
ax = plt.subplot(211)
ax.set_title('Normal Distributions')
g1, g2 = (x1, y1), (x2, y2)
data = (g1, g2)
colors = ('red', 'green')
groups = ('dataset1', 'dataset2')
for data, color, group in zip(data, colors, groups):
    x, y = data
    ax.scatter(x, y, alpha=0.8, c=color, edgecolors='none',
               s=30, label=group)
plt.legend(loc=4)
ax = plt.subplot(212)
ax.set_title('Standard Normal Distributions')
ds1 = (x1, std_nrml(y1, mu1, sigma1))
y1_sn = ds1[1]
ds2 = (x2, std_nrml(y2, mu2, sigma2))
y2_sn = ds2[1]
g1, g2 = (x1, y1_sn), (x2, y2_sn)
data = (g1, g2)
```

```
for data, color, group in zip(data, colors, groups):
    x, y = data
    ax.scatter(x, y, alpha=0.8, c=color, edgecolors='none',
               s=30, label=group)
plt.tight_layout()
plt.show()
```

Output:

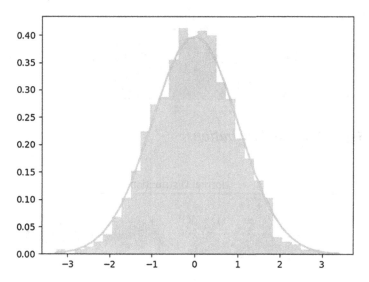

Figure 5-12. *Standard normal distribution*

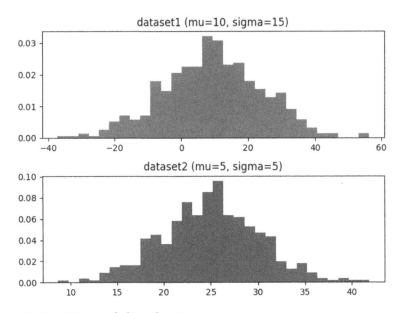

Figure 5-13. *Normal distributions*

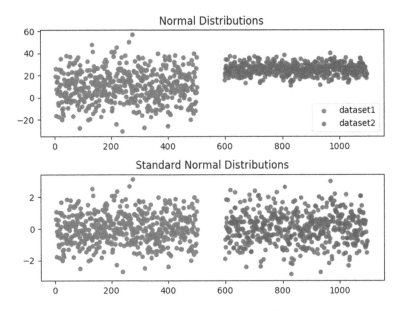

Figure 5-14. *Normal and standard normal distributions*

The code example begins by importing numpy and matplotlib. Function rnd_nrml() generates a normal distribution based on mean (mu), SD (sigma), and n number of data points. Function std_nrml() transforms data to standard normal. The main block begins by creating a standard normal distribution as a histogram and a line (Figure 5-12). The code continues by creating and plotting two different normally distributed datasets (Figure 5-13). Next, both data sets are rescaled to standard normal and plotted (Figure 5-14). Now, the datasets can be compared with each other. Although the original plots of the datasets appear to be very different, they are actually very similar distributions.

The 4th code example reads a CSV dataset, saves it to JSON, wrangles it, and prints a few records. The URL for the data is: https://community. tableau.com/docs/DOC-1236. However, the data on this site changes, so please use the data from our website to work with this example:

```
import csv, json

def read_dict(f):
    return csv.DictReader(open(f))

def to_dict(d):
    return [dict(row) for row in d]

def dump_json(f, d):
    with open(f, 'w') as fout:
        json.dump(d, fout)

def read_json(f):
    with open(f) as f:
        return json.load(f)
```

```python
def mk_data(d):
    for i, row in enumerate(d):
        e = {}
        e['_id'] = i
        e['cust'] = row['Customer Name']
        e['item'] = row['Sub-Category']
        e['sale'] = rnd(row['Sales'])
        e['quan'] = row['Quantity']
        e['disc'] = row['Discount']
        e['prof'] = rnd(row['Profit'])
        e['segm'] = row['Segment']
        yield e

def rnd(v):
    return str(round(float(v),2))

if __name__ == "__main__":
    f= 'data/superstore.csv'
    d = read_dict(f)
    data = to_dict(d)
    jsonf = 'data/superstore.json'
    dump_json(jsonf, data)
    print ('"superstore" data added to JSON\n')
    json_data = read_json(jsonf)
    print ("{:20s} {:15s} {:10s} {:3s} {:5s} {:12s} {:10s}".
            format('CUSTOMER', 'ITEM', 'SALES', 'Q', 'DISC',
                    'PROFIT', 'SEGMENT'))
    generator = mk_data(json_data)
    for i, row in enumerate(generator):
        if i < 10:
            print ("{:20s} {:15s}".format(row['cust'],
                    row['item'])),
```

```
            "{:10s} {:3s}".format(row['sale'],
            row['quan']),
            "{:5s} {:12s}".format(row['disc'],
            row['prof']),
            "{:10s}".format(row['segm']))
    else:
        break
```

Output:

```
"superstore" data added to JSON

CUSTOMER             ITEM          SALES     Q   DISC   PROFIT      SEGMENT
Claire Gute          Bookcases     261.96    2   0      41.91       Consumer
Claire Gute          Chairs        731.94    3   0      219.58      Consumer
Darrin Van Huff      Labels        14.62     2   0      6.87        Corporate
Sean O'Donnell       Tables        957.58    5   0.45   -383.03     Consumer
Sean O'Donnell       Storage       22.37     2   0.2    2.52        Consumer
Brosina Hoffman      Furnishings   48.86     7   0      14.17       Consumer
Brosina Hoffman      Art           7.28      4   0      1.97        Consumer
Brosina Hoffman      Phones        907.15    6   0.2    90.72       Consumer
Brosina Hoffman      Binders       18.5      3   0.2    5.78        Consumer
Brosina Hoffman      Appliances    114.9     5   0      34.47       Consumer
```

The code example begins by importing csv and json libraries. Function read_dict() reads a CSV file as an OrderedDict. Function to_dict() converts an OrderedDict to a regular dictionary. Function dump_json() saves a file to JSON. Function read_json() reads a JSON file. Function mk_data() creates a generator object consisting of wrangled data from the JSON file. Function rnd() rounds a number to 2 decimal places. The main block begins by reading a CSV file and converting it to JSON. The code continues by reading the newly created JSON data. Next, a generator object is created from the JSON data. The generator object is critical because it speeds processing orders of magnitude faster than a list. Since the dataset is close to 10,000 records, speed is important. To verify that the data was created correctly, the generator object is iterated a few times to print some of the wrangled records.

The 5th and final code example reads the JSON file created in the previous example, wrangles it, and saves the wrangled data set to JSON:

```python
import json

def read_json(f):
    with open(f) as f:
        return json.load(f)

def mk_data(d):
    for i, row in enumerate(d):
        e = {}
        e['_id'] = i
        e['cust'] = row['Customer Name']
        e['item'] = row['Sub-Category']
        e['sale'] = rnd(row['Sales'])
        e['quan'] = row['Quantity']
        e['disc'] = row['Discount']
        e['prof'] = rnd(row['Profit'])
        e['segm'] = row['Segment']
        yield e

def rnd(v):
    return str(round(float(v),2))

if __name__ == "__main__":
    jsonf = 'data/superstore.json'
    json_data = read_json(jsonf)
    l = len(list(mk_data(json_data)))
    generator = mk_data(json_data)
    jsonf= 'data/wrangled.json'
    with open(jsonf, 'w') as f:
        f.write('[')
    for i, row in enumerate(generator):
        j = json.dumps(row)
```

```
    if i < l - 1:
        with open(jsonf, 'a') as f:
            f.write(j)
            f.write(',')
    else:
        with open(jsonf, 'a') as f:
            f.write(j)
            f.write(']')
json_data = read_json(jsonf)
for i, row in enumerate(json_data):
    if i < 5:
        print (row['cust'], row['item'], row['sale'])
    else:
        break
```

Output:

```
Claire Gute Bookcases 261.96
Claire Gute Chairs 731.94
Darrin Van Huff Labels 14.62
Sean O'Donnell Tables 957.58
Sean O'Donnell Storage 22.37
```

The code example imports json. Function read_json() reads a JSON file. Function mk_data() creates a generator object consisting of wrangled data from the JSON file. Function rnd() rounds a number to two decimal places. The main block begins by reading a JSON file. A generator object must be created twice. The 1st generator allows us to find the length of the JSON file. The 2nd generator consists of wrangled data from the JSON file. Next, the generator is traversed so we can create a JSON file of the wrangled data. Although the generator object is created and can be traversed very fast, it takes a bit of time to create a JSON file consisting of close to 10,000 wrangled records. On my machine, it took a bit over 33 seconds, so be patient.

CHAPTER 6

Exploring Data

Exploring probes deeper into the realm of data. An important topic in data science is dimensionality reduction. This chapter borrows munged data from Chapter 5 to demonstrate how this works. Another topic is speed simulation. When working with large datasets, speed is of great importance. Big data is explored with a popular dataset used by academics and industry. Finally, Twitter and Web scraping are two important data sources for exploration.

Heat Maps

Heat maps were introduced in Chapter 5, but one wasn't created for the munged dataset. So, we start by creating a Heat map visualization of the wrangled.json data.

```
import json, pandas as pd
import matplotlib.pyplot as plt
import seaborn as sns

def read_json(f):
    with open(f) as f:
        return json.load(f)

def verify_keys(d, **kwargs):
    data = d[0].items()
    k1 = set([tup[0] for tup in data])
```

© David Paper 2018
D. Paper, *Data Science Fundamentals for Python and MongoDB*,
https://doi.org/10.1007/978-1-4842-3597-3_6

```
    s = kwargs.items()
    k2 = set([tup[1] for tup in s])
    return list(k1.intersection(k2))

def build_ls(k, d):
    return [{k: row[k] for k in (keys)} for row in d]

def get_rows(d, n):
    [print(row) for i, row in enumerate(d) if i < n]

def conv_float(d):
    return [dict([k, float(v)] for k, v in row.items()) for row
in d]

if __name__ == "__main__":
    f= 'data/wrangled.json'
    data = read_json(f)
    keys = verify_keys(data, c1='sale', c2='quan', c3='disc',
    c4='prof')
    heat = build_ls(keys, data)
    print ('1st row in "heat":')
    get_rows(heat, 1)
    heat = conv_float(heat)
    print ('\n1st row in "heat" converted to float:')
    get_rows(heat, 1)
    df = pd.DataFrame(heat)
    plt.figure()
    sns.heatmap(df.corr(), annot=True, cmap='OrRd')
    plt.show()
```

Output:

```
1st row in "heat":
{'prof': '41.91', 'disc': '0', 'quan': '2', 'sale': '261.96'}

1st row in "heat" converted to float:
{'prof': 41.91, 'disc': 0.0, 'quan': 2.0, 'sale': 261.96}
```

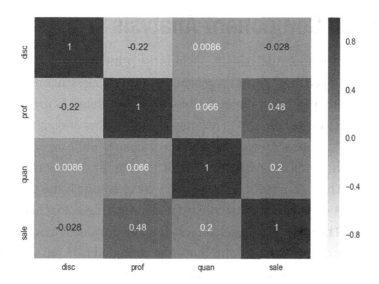

Figure 6-1. *Heat map*

The code example begins by importing json, pandas, matplotlib, and seaborn libraries. Function read_json() reads a JSON file. Function verify_keys() ensures that the keys of interest exist in the JSON file. This is important because we can only create a Heat map based on numerical variables, and the only candidates from the JSON file are sales, quantity, discount, and profit. Function build_ls() builds a list of dictionary elements based on the numerical variables. Function get_rows() returns n rows from a list. Function conv_float() converts dictionary elements to float. The main block begins by reading JSON file wrangled.json. It continues by getting keys for only numerical variables. Next, it builds list a list of dictionary elements (heat) based on the appropriate keys. The code displays the 1st row in heat to verify that all values are float. Since they are not, the code converts them to float. The code then creates a df from heat and plots the Heat map (Figure 6-1).

169

Principal Component Analysis

Principal Component Analysis (PCA) finds the principal components of data. Principal components represent the underlying structure in the data because they uncover the directions where the data has the most variance (most spread out). PCA leverages eigenvectors and eigenvalues to uncover data variance. An eigenvector is a direction, while an eigenvalue is a number that indicates variance (in the data) in the direction of the eigenvector. The eigenvector with the highest eigenvalue is the principal component. A dataset can be deconstructed into eigenvectors and eigenvalues. The amount of eigenvectors (and eigenvalues) in a dataset equals the number of dimensions. Since the wrangled.json dataset has four dimensions (variables), it has four eigenvectors/eigenvalues.

The 1st code example runs PCA on the wrangled.json dataset. However, PCA only works with numeric data, so the dataset is distilled down to only those features.

```
import matplotlib.pyplot as plt, pandas as pd
import numpy as np, json, random as rnd
from sklearn.preprocessing import StandardScaler
from pandas.plotting import parallel_coordinates

def read_json(f):
    with open(f) as f:
        return json.load(f)

def unique_features(k, d):
    return list(set([dic[k] for dic in d]))

def sire_features(k, d):
    return [{k: row[k] for k in (k)} for row in d]

def sire_numeric(k, d):
    s = conv_float(sire_features(k, d))
    return s
```

```python
def sire_sample(k, v, d, m):
    indices = np.arange(0, len(d), 1)
    s = [d[i] for i in indices if d[i][k] == v]
    n = len(s)
    num_keys = ['sale', 'quan', 'disc', 'prof']
    for i, row in enumerate(s):
        for k in num_keys:
            row[k] = float(row[k])
    s = rnd_sample(m, len(s), s)
    return (s, n)

def rnd_sample(m, n, d):
    indices = sorted(rnd.sample(range(n), m))
    return [d[i] for i in indices]

def conv_float(d):
    return [dict([k, float(v)] for k, v in row.items()) for row
    in d]

if __name__ == "__main__":
    f = 'data/wrangled.json'
    data = read_json(f)
    segm = unique_features('segm', data)
    print ('classes in "segm" feature:')
    print (segm)
    keys = ['sale', 'quan', 'disc', 'prof', 'segm']
    features = sire_features(keys, data)
    num_keys = ['sale', 'quan', 'disc', 'prof']
    numeric_data = sire_numeric(num_keys, features)
    k, v = "segm", "Home Office"
    m = 100
    s_home = sire_sample(k, v, features, m)
    v = "Consumer"
    s_cons = sire_sample(k, v, features, m)
```

```
v = "Corporate"
s_corp = sire_sample(k, v, features, m)
print ('\nHome Office slice:', s_home[1])
print('Consumer slice:', s_cons[1])
print ('Coporate slice:', s_corp[1])
print ('sample size:', m)
df_home = pd.DataFrame(s_home[0])
df_cons = pd.DataFrame(s_cons[0])
df_corp = pd.DataFrame(s_corp[0])
frames = [df_home, df_cons, df_corp]
result = pd.concat(frames)
plt.figure()
parallel_coordinates(result, 'segm', color=
                        ['orange','lime','fuchsia'])
df = pd.DataFrame(numeric_data)
X = df.ix[:].values
X_std = StandardScaler().fit_transform(X)
mean_vec = np.mean(X_std, axis=0)
cov_mat = np.cov(X_std.T)
print ('\ncovariance matrix:\n', cov_mat)
eig_vals, eig_vecs = np.linalg.eig(cov_mat)
print ('\nEigenvectors:\n', eig_vecs)
print ('\nEigenvalues:\n', np.sort(eig_vals)[::-1])
tot = sum(eig_vals)
var_exp = [(i / tot)*100 for i in sorted(eig_vals,
reverse=True)]
print ('\nvariance explained:\n', var_exp)
corr_mat = np.corrcoef(X.T)
print ('\ncorrelation matrix:\n', corr_mat)
eig_vals, eig_vecs = np.linalg.eig(corr_mat)
print ('\nEigenvectors:\n', eig_vecs)
print ('\nEigenvalues:\n', np.sort(eig_vals)[::-1])
```

```
tot = sum(eig_vals)
var_exp = [(i / tot)*100 for i in sorted(eig_vals,
reverse=True)]
print ('\nvariance explained:\n', var_exp)
cum_var_exp = np.cumsum(var_exp)
fig, ax = plt.subplots()
labels = ['PC1', 'PC2', 'PC3', 'PC4']
width = 0.35
index = np.arange(len(var_exp))
ax.bar(index, var_exp,
       color=['fuchsia', 'lime', 'thistle', 'thistle'])
for i, v in enumerate(var_exp):
    v = round(v, 2)
    val = str(v) + '%'
    ax.text(i, v+0.5, val, ha='center', color='b',
            fontsize=9, fontweight='bold')
plt.xticks(index, labels)
plt.title('Variance Explained')
plt.show()
```

Output:

```
classes in "segm" feature:
['Home Office', 'Consumer', 'Corporate']

Home Office slice: 1783
Consumer slice: 5191
Coporate slice: 3020
sample size: 100
```

```
covariance matrix:
 [[ 1.00010007 -0.21950937  0.00862383 -0.02819299]
 [-0.21950937  1.00010007  0.06625978  0.47911246]
 [ 0.00862383  0.06625978  1.00010007  0.20081494]
 [-0.02819299  0.47911246  0.20081494  1.00010007]]

Eigenvectors:
 [[-0.27037624  0.24517839  0.71856986  0.59198108]
 [ 0.65545599  0.68416982 -0.19540197  0.25319396]
 [ 0.28863648  0.16325021  0.63082196 -0.70149983]
 [ 0.64339966 -0.66719456  0.21803458  0.30553103]]

Eigenvalues:
 [ 1.59012581  1.05880782  0.88144442  0.47002223]

variance explained:
 [39.749167611521841, 26.467546859432439, 22.033905616069589, 11.749379912976137]

correlation matrix:
 [[ 1.         -0.21948741  0.00862297 -0.02819017]
 [-0.21948741  1.          0.06625315  0.47906452]
 [ 0.00862297  0.06625315  1.          0.20079484]
 [-0.02819017  0.47906452  0.20079484  1.        ]]

Eigenvectors:
 [[-0.27037624  0.24517839  0.71856986  0.59198108]
 [ 0.65545599  0.68416982 -0.19540197  0.25319396]
 [ 0.28863648  0.16325021  0.63082196 -0.70149983]
 [ 0.64339966 -0.66719456  0.21803458  0.30553103]]

Eigenvalues:
 [ 1.5899667   1.05870187  0.88135622  0.4699752 ]

variance explained:
 [39.749167611521855, 26.467546859432396, 22.033905616069603, 11.749379912976144]
```

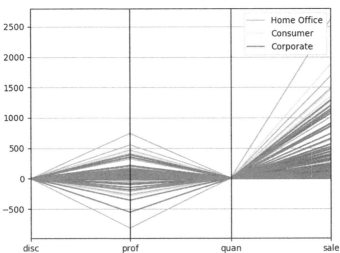

Figure 6-2. *Parallel coordinates*

174

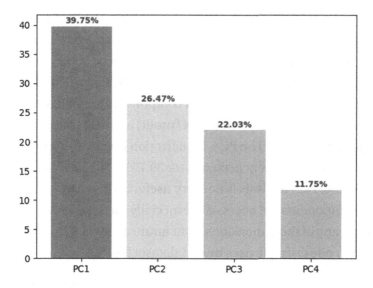

Figure 6-3. *Variance explained*

The code example begins by importing matplotlib, pandas, numpy, json, random, and sklearn libraries. Function read_json() reads a JSON file. Function unique_features() distills unique categories (classes) from a dimension (feature). In this case, it distills three classes—Home Office, Corporate, and Consumer—from the segm feature. Since the dataset is close to 10,000 records, I wanted to be sure what classes are in it. Function sire_features() distills a new dataset with only features of interest. Function sire_numeric() converts numeric strings to float. Function sire_sample() returns a random sample of n records filtered for a class. Function rnd_sample() creates a random sample. Function convert_float() converts numeric string data to float.

The main block begins by reading wrangled.json and creating dataset features with only features of interest. The code continues by creating dataset numeric that only includes features with numeric data. Dataset numeric is used to generate PCA. Next, three samples of size 100 are created; one for each class. The samples are used to create the

175

parallel coordinates visualization (Figure 6-2). Code for PCA follows by standardizing and transforming the numeric dataset. A covariance matrix is created so that eigenvectors and eigenvalues can be generated. I include PCA using the correlation matrix because some disciplines prefer it. Finally, a visualization of the principal components is created.

Parallel coordinates show that prof (profit) and sale (sales) are the most important features. The PCA visualization (Figure 6-3) shows that the 1st principal component accounts for 39.75%, 2nd 26.47%, 3rd 22.03%, and 4th 11.75%. PCA analysis is not very useful in this case, since all four principal components are necessary, especially the 1st three. So, we cannot drop any of the dimensions from future analysis.

The 2nd code example uses the iris dataset for PCA:

```
import matplotlib.pyplot as plt, pandas as pd, numpy as np
from sklearn.preprocessing import StandardScaler
from pandas.plotting import parallel_coordinates

def conv_float(d):
    return d.astype(float)

if __name__ == "__main__":
    df = pd.read_csv('data/iris.csv')
    X = df.ix[:,0:4].values
    y = df.ix[:,4].values
    X_std = StandardScaler().fit_transform(X)
    mean_vec = np.mean(X_std, axis=0)
    cov_mat = np.cov(X_std.T)
    eig_vals, eig_vecs = np.linalg.eig(cov_mat)
    print ('Eigenvectors:\n', eig_vecs)
    print ('\nEigenvalues:\n', eig_vals)
    plt.figure()
    parallel_coordinates(df, 'Name', color=
                            ['orange','lime','fuchsia'])
```

```
tot = sum(eig_vals)
var_exp = [(i / tot)*100 for i in sorted(eig_vals,
reverse=True)]
cum_var_exp = np.cumsum(var_exp)
fig, ax = plt.subplots()
labels = ['PC1', 'PC2', 'PC3', 'PC4']
width = 0.35
index = np.arange(len(var_exp))
ax.bar(index, var_exp,
        color=['fuchsia', 'lime', 'thistle', 'thistle'])
for i, v in enumerate(var_exp):
    v = round(v, 2)
    val = str(v) + '%'
    ax.text(i, v+0.5, val, ha='center', color='b',
            fontsize=9, fontweight='bold')
plt.xticks(index, labels)
plt.title('Variance Explained')
plt.show()
```

Output:

```
Eigenvectors:
 [[ 0.52237162 -0.37231836 -0.72101681  0.26199559]
  [-0.26335492 -0.92555649  0.24203288 -0.12413481]
  [ 0.58125401 -0.02109478  0.14089226 -0.80115427]
  [ 0.56561105 -0.06541577  0.6338014   0.52354627]]

Eigenvalues:
 [ 2.93035378  0.92740362  0.14834223  0.02074601]
```

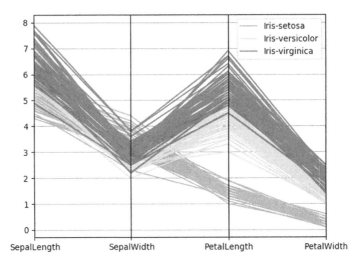

Figure 6-4. *Parallel coordinates*

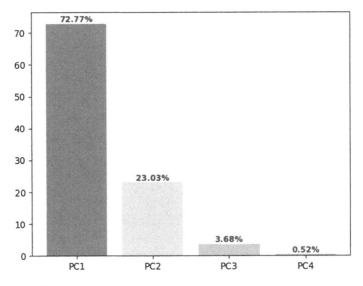

Figure 6-5. *Variance explained*

The code example is much shorter than the previous one, because we didn't have to wrangle, clean (as much), and create random samples (for Parallel Coordinates visualization). The code begins by importing matplotlib, pandas, numpy, and sklearn libraries. Function conv_float() converts numeric strings to float. The main block begins by reading the iris dataset. It continues by standardizing and transforming the data for PCA. Parallel Coordinates and variance explained are then displayed.

Parallel Coordinates shows that PetalLength and PetalWidth are the most important features (Figure 6-4). The PCA visualization (Variance Explained) shows that the 1st principal component accounts for 72.77%, 2nd 23.03%, 3rd 3.68%, and 4th 0.52% (Figure 6-5). PCA analysis is very useful in this case because the 1st two principal components account for over 95% of the variance. So, we can drop PC3 and PC4 from further consideration.

For clarity, the 1st step for PCA is to explore the eigenvectors and eigenvalues. The eigenvectors with the lowest eigenvalues bear the least information about the distribution of the data, so they can be dropped. In this example, the 1st two eigenvalues are much higher, especially PC1. Dropping PC3 and PC4 are thereby in order. The 2nd step is to measure explained variance, which can be calculated from the eigenvalues. Explained variance tells us how much information (variance) can be attributed to each of the principal components. Looking at explained variance confirms that PC3 and PC4 are not important.

Speed Simulation

Speed in data science is important, especially as datasets become bigger. Generators are helpful in memory optimization, because a generator function returns one item at a time (as needed) rather than all items at once.

The code example contrasts speed between a list and a generator:

```python
import json, humanfriendly as hf
from time import clock

def read_json(f):
    with open(f) as f:
        return json.load(f)

def mk_gen(k, d):
    for row in d:
        dic = {}
        for key in k:
            dic[key] = float(row[key])
        yield dic

def conv_float(keys, d):
    return [dict([k, float(v)] for k, v in row.items()
                 if k in keys) for row in d]

if __name__ == "__main__":
    f = 'data/wrangled.json'
    data = read_json(f)
    keys = ['sale', 'quan', 'disc', 'prof']
    print ('create, convert, and display list:')
    start = clock()
    data = conv_float(keys, data)
    for i, row in enumerate(data):
        if i < 5:
            print (row)
    end = clock()
    elapsed_ls = end - start
    print (hf.format_timespan(elapsed_ls, detailed=True))
    print ('\ncreate, convert, and display generator:')
```

```
start = clock()
generator = mk_gen(keys, data)
for i, row in enumerate(generator):
    if i < 5:
        print (row)
end = clock()
elapsed_gen = end - start
print (hf.format_timespan(elapsed_gen, detailed=True))
speed = round(elapsed_ls / elapsed_gen, 2)
print ('\ngenerator is', speed, 'times faster')
```

Output:

```
create, convert, and display list:
{'sale': 261.96, 'quan': 2.0, 'disc': 0.0, 'prof': 41.91}
{'sale': 731.94, 'quan': 3.0, 'disc': 0.0, 'prof': 219.58}
{'sale': 14.62, 'quan': 2.0, 'disc': 0.0, 'prof': 6.87}
{'sale': 957.58, 'quan': 5.0, 'disc': 0.45, 'prof': -383.03}
{'sale': 22.37, 'quan': 2.0, 'disc': 0.2, 'prof': 2.52}
46.03 milliseconds

create, convert, and display generator:
{'sale': 261.96, 'quan': 2.0, 'disc': 0.0, 'prof': 41.91}
{'sale': 731.94, 'quan': 3.0, 'disc': 0.0, 'prof': 219.58}
{'sale': 14.62, 'quan': 2.0, 'disc': 0.0, 'prof': 6.87}
{'sale': 957.58, 'quan': 5.0, 'disc': 0.45, 'prof': -383.03}
{'sale': 22.37, 'quan': 2.0, 'disc': 0.2, 'prof': 2.52}
20.38 milliseconds

generator is 2.26 times faster
```

The code example begins by importing json, humanfriendly, and time libraries. You may have to install humanfriendly like I did as so: *pip install humanfriendly*. Function read_json() reads JSON. Function mk_gen() creates a generator based on four features from wrangled.json and converts values to float. Function conv_float() converts dictionary values from a list to float. The main block begins by reading wrangled. json into a list. The code continues by timing the process of creating a new list from keys and converting values to float. Next, a generator is created that mimics the list creating and conversion process. The generator is 2.26 times faster (on my computer).

Big Data

Big data is the rage of the 21st century. So, let's work with a relatively big dataset. GroupLens is a website that offers access to large social computing datasets for theory and practice. GroupLens has collected and made available rating datasets from the MovieLens website:

https://grouplens.org/datasets/movielens/. We are going to explore the 1M dataset, which contains approximately one million ratings from six thousand users on four thousand movies. I was hesitant to wrangle, cleanse, and process a dataset over one million because of the limited processing power of my relatively new PC.

The 1st code example reads, cleans, sizes, and dumps MovieLens data to JSON:

```
import json, csv

def read_dat(h, f):
    return csv.DictReader((line.replace('::', ':')
                            for line in open(f)),
                           delimiter=':', fieldnames=h,
                           quoting=csv.QUOTE_NONE)

def gen_dict(d):
    for row in d:
        yield dict(row)

def dump_json(f, l, d):
    f = open(f, 'w')
    f.write('[')
    for i, row in enumerate(d):
        j = json.dumps(row)
        f.write(j)
        if i < l - 1:
            f.write(',')
```

```
        else:
            f.write(']')
    f.close()

def read_json(f):
    with open(f) as f:
        return json.load(f)

def display(n, f):
    for i, row in enumerate(f):
        if i < n:
            print (row)
    print()

if __name__ == "__main__":
    print ('... sizing data ...\n')
    u_dat = 'data/ml-1m/users.dat'
    m_dat = 'data/ml-1m/movies.dat'
    r_dat = 'data/ml-1m/ratings.dat'
    unames = ['user_id', 'gender', 'age', 'occupation', 'zip']
    mnames = ['movie_id', 'title', 'genres']
    rnames = ['user_id', 'movie_id', 'rating', 'timestamp']
    users = read_dat(unames, u_dat)
    ul = len(list(gen_dict(users)))
    movies = read_dat(mnames, m_dat)
    ml = len(list(gen_dict(movies)))
    ratings = read_dat(rnames, r_dat)
    rl = len(list(gen_dict(ratings)))
    print ('size of datasets:')
    print ('users', ul)
    print ('movies', ml)
    print ('ratings', rl)
    print ('\n... dumping data ...\n')
```

```
users = read_dat(unames, u_dat)
users = gen_dict(users)
movies = read_dat(mnames, m_dat)
movies = gen_dict(movies)
ratings = read_dat(rnames, r_dat)
ratings = gen_dict(ratings)
uf = 'data/users.json'
dump_json(uf, ul, users)
mf = 'data/movies.json'
dump_json(mf, ml, movies)
rf = 'data/ratings.json'
dump_json(rf, rl, ratings)
print ('\n... verifying data ...\n')
u = read_json(uf)
m = read_json(mf)
r = read_json(rf)
n = 1
display(n, u)
display(n, m)
display(n, r)
```

Output:

```
... sizing data ...

size of datasets:
users 6040
movies 3883
ratings 1000209

... dumping data ...

... verifying data ...

{'user_id': '1', 'gender': 'F', 'age': '1', 'occupation': '10', 'zip': '48067'}

{'movie_id': '1', 'title': 'Toy Story (1995)', 'genres': "Animation|Children's|Comedy"}

{'user_id': '1', 'movie_id': '1193', 'rating': '5', 'timestamp': '978300760'}
```

The code example begins by importing json and csv libraries. Function read_dat() reads and cleans the data (replaces double colons with single colons as delimiters). Function gen_dict() converts an OrderedDict list to a regular dictionary list for easier processing. Function dump_json() is a custom function that I wrote to dump data to JSON. Function read_json() reads JSON. Function display() displays some data for verification. The main block begins by reading the three datasets and finding their sizes. It continues by rereading the datasets and dumping to JSON. The datasets need to be reread, because a generator can only be traversed once. Since the ratings dataset is over one million records, it takes a few seconds to process.

The 2nd code example cleans the movie dataset, which requires extensive additional cleaning:

```python
import json, numpy as np

def read_json(f):
    with open(f) as f:
        return json.load(f)

def dump_json(f, d):
    with open(f, 'w') as fout:
        json.dump(d, fout)

def display(n, d):
    [print (row) for i,row in enumerate(d) if i < n]

def get_indx(k, d):
    return [row[k] for row in d if 'null' in row]

def get_data(k, l, d):
    return [row for i, row in enumerate(d) if row[k] in l]

def get_unique(key, d):
    s = set()
    for row in d:
```

```
            for k, v in row.items():
                if k in key:
                    s.add(v)
        return np.sort(list(s))

if __name__ == "__main__":
    mf = 'data/movies.json'
    m = read_json(mf)
    n = 20
    display(n, m)
    print ()
    indx = get_indx('movie_id', m)
    for row in m:
        if row['movie_id'] in indx:
            row['title'] = row['title'] + ':' + row['genres']
            row['genres'] = row['null'][0]
            del row['null']
        title = row['title'].split(" ")
        year = title.pop()
        year = ''.join(c for c in year if c not in '()')
        row['title'] = ' '.join(title)
        row['year'] = year
    data = get_data('movie_id', indx, m)
    n = 2
    display(n, data)
    s = get_unique('year', m)
    print ('\n', s, '\n')
    rec = get_data('year', ['Assignment'], m)
    print (rec[0])
```

```
rec = get_data('year', ["L'Associe1982"], m)
print (rec[0], '\n')
b1, b2, cnt = False, False, 0
for row in m:
    if row['movie_id'] in ['1001']:
        row['year'] = '1982'
        print (row)
        b1 = True
    elif row['movie_id'] in ['2382']:
        row['title'] = 'Police Academy 5: Assignment: Miami
        Beach'
        row['genres'] = 'Comedy'
        row['year'] = '1988'
        print (row)
        b2 = True
    elif b1 and b2: break
    cnt += 1
print ('\n', cnt, len(m))
mf = 'data/cmovies.json'
dump_json(mf, m)
m = read_json(mf)
display(n, m)
```

Output:

```
{'movie_id': '1', 'title': 'Toy Story (1995)', 'genres': "Animation|Children's|Comedy"}
{'movie_id': '2', 'title': 'Jumanji (1995)', 'genres': "Adventure|Children's|Fantasy"}
{'movie_id': '3', 'title': 'Grumpier Old Men (1995)', 'genres': 'Comedy|Romance'}
{'movie_id': '4', 'title': 'Waiting to Exhale (1995)', 'genres': 'Comedy|Drama'}
{'movie_id': '5', 'title': 'Father of the Bride Part II (1995)', 'genres': 'Comedy'}
{'movie_id': '6', 'title': 'Heat (1995)', 'genres': 'Action|Crime|Thriller'}
{'movie_id': '7', 'title': 'Sabrina (1995)', 'genres': 'Comedy|Romance'}
{'movie_id': '8', 'title': 'Tom and Huck (1995)', 'genres': "Adventure|Children's"}
{'movie_id': '9', 'title': 'Sudden Death (1995)', 'genres': 'Action'}
{'movie_id': '10', 'title': 'GoldenEye (1995)', 'genres': 'Action|Adventure|Thriller'}
{'movie_id': '11', 'title': 'American President, The (1995)', 'genres': 'Comedy|Drama|Romance'}
{'movie_id': '12', 'title': 'Dracula', 'genres': ' Dead and Loving It (1995)', 'null': ['Comedy|Horror']}
{'movie_id': '13', 'title': 'Balto (1995)', 'genres': "Animation|Children's"}
{'movie_id': '14', 'title': 'Nixon (1995)', 'genres': 'Drama'}
{'movie_id': '15', 'title': 'Cutthroat Island (1995)', 'genres': 'Action|Adventure|Romance'}
{'movie_id': '16', 'title': 'Casino (1995)', 'genres': 'Drama|Thriller'}
{'movie_id': '17', 'title': 'Sense and Sensibility (1995)', 'genres': 'Drama|Romance'}
{'movie_id': '18', 'title': 'Four Rooms (1995)', 'genres': 'Thriller'}
{'movie_id': '19', 'title': 'Ace Ventura', 'genres': ' When Nature Calls (1995)', 'null': ['Comedy']}
{'movie_id': '20', 'title': 'Money Train (1995)', 'genres': 'Action'}

{'movie_id': '12', 'title': 'Dracula: Dead and Loving It', 'genres': 'Comedy|Horror', 'year': '1995'}
{'movie_id': '19', 'title': 'Ace Ventura: When Nature Calls', 'genres': 'Comedy', 'year': '1995'}

['1919' '1920' '1921' '1922' '1923' '1925' '1926' '1927' '1928' '1929'
 '1930' '1931' '1932' '1933' '1934' '1935' '1936' '1937' '1938' '1939'
 '1940' '1941' '1942' '1943' '1944' '1945' '1946' '1947' '1948' '1949'
 '1950' '1951' '1952' '1953' '1954' '1955' '1956' '1957' '1958' '1959'
 '1960' '1961' '1962' '1963' '1964' '1965' '1966' '1967' '1968' '1969'
 '1970' '1971' '1972' '1973' '1974' '1975' '1976' '1977' '1978' '1979'
 '1980' '1981' '1982' '1983' '1984' '1985' '1986' '1987' '1988' '1989'
 '1990' '1991' '1992' '1993' '1994' '1995' '1996' '1997' '1998' '1999'
 '2000' 'Assignment' "L'Associe1982"]

{'movie_id': '2382', 'title': 'Police Academy 5:', 'genres': ' Miami Beach (1988)', 'year': 'Assignment'}
{'movie_id': '1001', 'title': 'Associate, The', 'genres': 'Comedy', 'year': "L'Associe1982"}

{'movie_id': '1001', 'title': 'Associate, The', 'genres': 'Comedy', 'year': '1982'}
{'movie_id': '2382', 'title': 'Police Academy 5: Assignment: Miami Beach', 'genres': 'Comedy', 'year': '1988'}

2314 3883
{'movie_id': '1', 'title': 'Toy Story', 'genres': "Animation|Children's|Comedy", 'year': '1995'}
{'movie_id': '2', 'title': 'Jumanji', 'genres': "Adventure|Children's|Fantasy", 'year': '1995'}
```

The code example begins by importing json and numpy libraries.
Function read_json() reads JSON. Function dump_json() saves
JSON. Function display() displays n records. Function get_indx() returns
indices of dictionary elements with a null key. Function get_data() returns
a dataset filtered by indices and movie_id key. Function get_unique()
returns a list of unique values from a list of dictionary elements. The main
block begins by reading movies.json and displaying for inspection. Records
12 and 19 have a null key. The code continues by finding all movie_id
indices with a null key. The next several lines clean all movies. Those with
a null key require added logic to fully clean, but all records have modified
titles and a new year key. To verify, records 12 and 19 are displayed.
To be sure that all is well, the code finds all unique keys based on year.

Notice that there are two records that don't have a legitimate year. So, the code cleans the two records. The 2nd elif was added to the code to stop processing once the two dirty records were cleaned. Although not included in the code, I checked movie_id, title, and genres keys but found no issues.

The code to connect to MongoDB is as follows:

```python
class conn:
    from pymongo import MongoClient
    client = MongoClient('localhost', port=27017)
    def __init__(self, dbname):
        self.db = conn.client[dbname]
    def getDB(self):
        return self.db
```

I created directory 'classes' and saved the code in 'conn.py'

The 3rd code example generates useful information from the three datasets:

```python
import json, numpy as np, sys, os, humanfriendly as hf
from time import clock
sys.path.append(os.getcwd()+'/classes')
import conn

def read_json(f):
    with open(f) as f:
        return json.load(f)

def get_column(A, v):
    return [A_i[v] for A_i in A]

def remove_nr(v1, v2):
    set_v1 = set(v1)
    set_v2 = set(v2)
    diff = list(set_v1 - set_v2)
    return diff
```

```python
def get_info(*args):
    a = [arg for arg in args]
    ratings = [int(row[a[0][1]]) for row in a[2] if row[a[0]
    [0]] == a[1]]
    uids = [row[a[0][3]] for row in a[2] if row[a[0][0]] == a[1]]
    title = [row[a[0][2]] for row in a[3] if row[a[0][0]] == a[1]]
    age = [int(row[a[0][4]]) for col in uids for row in a[4] if
    col == row[a[0][3]]]
    gender = [row[a[0][5]] for col in uids for row in users if
    col == row[a[0][3]]]
    return (ratings, title[0], uids, age, gender)

def generate(k, v, r, m, u):
    for i, mid in enumerate(v):
        dic = {}
        rec = get_info(k, mid, r, m, u)
        dic = {'_id':i, 'mid':mid, 'title':rec[1], 'avg_
        rating':np.mean(rec[0]),
                'n_ratings':len(rec[0]), 'avg_age':np.
                mean(rec[3]),
                'M':rec[4].count('M'), 'F':rec[4].count('F')}
        dic['avg_rating'] = round(float(str(dic['avg_rating'])
        [:6]),2)
        dic['avg_age'] = round(float(str(dic['avg_age'])[:6]))
        yield dic

def gen_ls(g):
    for i, row in enumerate(g):
        yield row
```

```python
if __name__ == "__main__":
    print ('... creating datasets ...\n')
    m = 'data/cmovies.json'
    movies = np.array(read_json(m))
    r = 'data/ratings.json'
    ratings = np.array(read_json(r))
    r = 'data/users.json'
    users = np.array(read_json(r))
    print ('... creating movie indicies vector data ...\n')
    mv = get_column(movies, 'movie_id')
    rv = get_column(ratings, 'movie_id')
    print ('... creating unrated movie indicies vector ...\n')
    nrv = remove_nr(mv, rv)
    diff = [int(row) for row in nrv]
    print (np.sort(diff), '\n')
    new_mv = [x for x in mv if x not in nrv]
    mid = '1'
    keys = ('movie_id', 'rating', 'title', 'user_id', 'age',
    'gender')
    stats = get_info(keys, mid, ratings, movies, users)
    avg_rating = np.mean(stats[0])
    avg_age = np.mean(stats[3])
    n_ratings = len(stats[0])
    title = stats[1]
    M, F = stats[4].count('M'), stats[4].count('F')
    print ('avg rating for:', end=' "')
    print (title + '" is', round(avg_rating, 2), end=' (')
    print (n_ratings, 'ratings)\n')
    gen = generate(keys, new_mv, ratings, movies, users)
    gls = gen_ls(gen)
    obj = conn.conn('test')
```

```
db = obj.getDB()
movie_info = db.movie_info
movie_info.drop()
print ('... saving movie_info to MongoDB ...\n')
start = clock()
for row in gls:
    movie_info.insert(row)
end = clock()
elapsed_ls = end - start
print (hf.format_timespan(elapsed_ls, detailed=True))
```

Output:

```
... creating datasets ...

... creating movie indicies vector data ...

... creating unrated movie indicies vector ...

[   51  109  115  143  284  285  395  399  400  403  604  620  625  629  636
   654  675  676  683  693  699  713  721  723  727  738  739  752  768  770
   772  773  777  794  795  797  812  816  819  822  825  845  855  856  857
   871  873  890  894  979  983 1001 1045 1052 1065 1075 1106 1108 1109 1110
  1122 1137 1140 1141 1143 1146 1155 1156 1157 1158 1159 1166 1308 1309 1314
  1318 1319 1368 1400 1424 1443 1448 1462 1467 1524 1557 1559 1568 1577 1578
  1628 1697 1698 1705 1706 1708 1710 1716 1723 1738 1740 1742 1757 1765 1768
  1774 1776 1781 1789 1819 1847 2030 2199 2216 2220 2222 2224 2225 2228 2229
  2230 2270 2274 2319 2489 2508 2547 2564 2588 2595 2601 2603 2604 2680 2684
  2698 2832 2838 2910 2954 2957 2958 2980 3009 3023 3059 3080 3170 3191 3193
  3195 3226 3227 3231 3234 3278 3279 3332 3348 3356 3369 3383 3411 3455 3541
  3558 3560 3561 3582 3583 3589 3630 3650 3750 3829 3856 3907]

avg rating for: "Toy Story" is 4.15 (2077 ratings)

... saving movie_info to MongoDB ...

31 minutes, 29 seconds and 96.07 milliseconds
```

The code example begins by importing json, numpy, sys, os, humanfriendly, time, and conn (a custom class I created to connect to MongoDB). Function read_json() reads JSON. Function get_column() returns a column vector. Function remove_nr() removes movie_id values that are not rated. Function get_info() returns ratings, users, age, and gender as column vectors as well as title of a movie. The function is very complex, because each vector is created by traversing one of the data sets and making comparisons. To make it more concise, list comprehension was used extensively. Function generate() generates a dictionary element that contains average rating, average age, number of males and females raters, number of ratings, movie_id, and title of each movie. Function gen_ ls() generates each dictionary element generated by function generate(). The main block begins by reading the three JSON datasets. It continues by getting two column vectors–each movie_id from movies dataset and movie_id from ratings dataset. Each column vector is converted to a set to remove duplicates. Column vectors are used instead of full records for faster processing. Next, a new column vector is returned containing only movies that are rated. The code continues by getting title and column vectors for ratings, and users, age, and gender for each movie with movie_ id of 1. The average rating for this movie is displayed with its title and number of ratings. The final part of the code creates a generator containing a list of dictionary elements. Each dictionary element contains the movie_ id, title, average rating, average age, number of ratings, number of male raters, and number of female raters. Next, another generator is created to generate the list. Creating the generators is instantaneous, but unraveling (unfolding) contents takes time. Keep in mind that the 1st generator runs billions of processes and 2nd generator runs the 1st one. So, saving contents to MongoDB takes close to half an hour.

To verify results, let's look at the data in MongoDB. The command show collections is the 1st that I run to check if collection movie_info was created:

```
> show collections
movie_info
```

Next, I run db.movie_info.count() to check the number of documents:

```
> db.movie_info.count()
3706
```

Now that I know the number of documents, I can display the first and last five records:

> db.movie_info.find({}, {title:1}).limit(5)

```
{'_id': 0, 'title': 'Toy Story'}
{'_id': 1, 'title': 'Jumanji'}
{'_id': 2, 'title': 'Grumpier Old Men'}
{'_id': 3, 'title': 'Waiting to Exhale'}
{'_id': 4, 'title': 'Father of the Bride Part II'}
```

> db.movie_info.find({}, {title:1}).skip(3701)

```
{'_id': 3701, 'title': 'Meet the Parents'}
{'_id': 3702, 'title': 'Requiem for a Dream'}
{'_id': 3703, 'title': 'Tigerland'}
{'_id': 3704, 'title': 'Two Family House'}
{'_id': 3705, 'title': 'Contender, The'}
```

From data exploration, it appears that the movie_info collection was created correctly.

The 4th code example saves the three datasets—users.json, cmovies. json, and ratings.json—to MongoDB:

```
import sys, os, json, humanfriendly as hf
from time import clock
sys.path.append(os.getcwd() + '/classes')
import conn

def read_json(f):
    with open(f) as f:
        return json.load(f)

def create_db(c, d):
    c = db[c]
    c.drop()
```

```
    for i, row in enumerate(d):
        row['_id'] = i
        c.insert(row)

if __name__ == "__main__":
    u = read_json('data/users.json')
    m = read_json('data/cmovies.json')
    r = read_json('data/ratings.json')
    obj = conn.conn('test')
    db = obj.getDB()
    print ('... creating MongoDB collections ...\n')
    start = clock()
    create_db('users', u)
    create_db('movies', m)
    create_db('ratings', r)
    end = clock()
    elapsed_ls = end - start
    print (hf.format_timespan(elapsed_ls, detailed=True))
```

Output:

```
... creating MongoDB collections ...

2 minutes, 28 seconds and 619.93 milliseconds
```

The code example begins by importing sys, os, json, humanfriendly, time, and custom class conn. Function read_json reads JSON. Function create_db() creates MongoDB collections. The main block begins by reading the three datasets–users.json, cmovies.json, and ratings.json–and saving them to MongoDB collections. Since the ratings.json dataset is over one million records, it takes some time to save it to the database.

The 5th code example introduces the aggregation pipeline, which is a MongoDB framework for data aggregation modeled on the concept of data processing pipelines. Documents enter a multistage pipeline that transforms them into aggregated results. In addition to grouping and sorting documents by specific field or fields and aggregating contents of arrays, pipeline stages can use operators for tasks such as calculating averages or concatenating strings. The pipeline provides efficient data aggregation using native MongoDB operations, and is the preferred method for data aggregation in MongoDB.

```python
import sys, os
sys.path.append(os.getcwd() + '/classes')
import conn

def match_item(k, v, d):
    pipeline = [ {'$match' : { k : v }} ]
    q = db.command('aggregate',d,pipeline=pipeline)
    return q

if __name__ == "__main__":
    obj = conn.conn('test')
    db = obj.getDB()
    movie = 'Toy Story'
    q = match_item('title', movie, 'movie_info')
    r = q['result'][0]
    print (movie, 'document:')
    print (r)
    print ('average rating', r['avg_rating'], '\n')
    user_id = '3'
    print ('*** user', user_id, '***')
    q = match_item('user_id', user_id, 'users')
    r = q['result'][0]
```

```python
print ('age', r['age'], 'gender', r['gender'],
'occupation',\
        r['occupation'], 'zip', r['zip'], '\n')
print ('*** "user 3" movie ratings of 5 ***')
q = match_item('user_id', user_id, 'ratings')
mid = q['result']
for row in mid:
    if row['rating'] == '5':
        q = match_item('movie_id', row['movie_id'], 'movies')
        title = q['result'][0]['title']
        genre = q['result'][0]['genres']
        print (row['movie_id'], title, genre)
mid = '1136'
q = match_item('mid', mid, 'movie_info')
title = q['result'][0]['title']
avg_rating = q['result'][0]['avg_rating']
print ()
print ('"' + title + '"', 'average rating:', avg_rating)
```

Output:

```
Toy Story document:
{'_id': 0, 'mid': '1', 'title': 'Toy Story', 'avg_rating': 4.15, 'n_ratings': 2077, 'avg_age': 28, 'M': 1486, 'F': 591}
average rating 4.15

*** user 3 ***
age 25 gender M occupation 15 zip 55117

*** "user 3" movie ratings of 5 ***
1079 Fish Called Wanda, A Comedy
1615 Edge, The Adventure|Thriller
1259 Stand by Me Adventure|Comedy|Drama
2167 Blade Action|Adventure|Horror
260 Star Wars: Episode IV - A New Hope Action|Adventure|Fantasy|Sci-Fi
1266 Unforgiven Western
733 Rock, The Action|Adventure|Thriller
2355 Bug's Life, A Animation|Children's|Comedy
1197 Princess Bride, The Action|Adventure|Comedy|Romance
1198 Raiders of the Lost Ark Action|Adventure
1378 Young Guns Action|Comedy|Western
3552 Caddyshack Comedy
1304 Butch Cassidy and the Sundance Kid Action|Comedy|Western
3671 Blazing Saddles Comedy|Western
1136 Monty Python and the Holy Grail Comedy

"Monty Python and the Holy Grail" average rating: 4.34
```

The code example begins by importing sys, os, and custom class conn. Function match_item() uses the aggregation pipeline to match records to criteria. The main block begins by using the aggregation pipeline to return the Toy Story document from collection movie_info. The code continues by using the pipeline to return the user 3 document from collection users. Next, the aggregation pipeline is used to return all movie ratings of 5 for user 3. Finally, the pipeline is used to return the average rating for Monty Python and the Holy Grail from collection movie_info. The aggregation pipeline is efficient and offers a vast array of functionality.

The 6th code example demonstrates a multistage aggregation pipeline:

```
import sys, os
sys.path.append(os.getcwd() + '/classes')
import conn

def stages(k, v, r, d):
    pipeline = [ {'$match' : { '$and' : [ { k : v },
                    {'rating':{'$eq':r} }] } },
                {'$project' : {
                    '_id' : 1,
                    'user_id' : 1,
                    'movie_id' : 1,
                    'rating' : 1 } },
                {'$limit' : 100}]
    q = db.command('aggregate',d,pipeline=pipeline)
    return q
```

```
def match_item(k, v, d):
    pipeline = [ {'$match' : { k : v }} ]
    q = db.command('aggregate',d,pipeline=pipeline)
    return q

if __name__ == "__main__":
    obj = conn.conn('test')
    db = obj.getDB()
    u = '3'
    r = '5'
    q = stages('user_id', u, r, 'ratings')
    result = q['result']
    print ('ratings of', r, 'for user ' + str(u) + ':')
    for i, row in enumerate(result):
        print (row)
    n = i+1
    print ()
    print (n, 'associated movie titles:')
    for i, row in enumerate(result):
        q = match_item('movie_id', row['movie_id'], 'movies')
        r = q['result'][0]
        print (r['title'])
```

Output:

```
ratings of 5 for user 3:
{'_id': 192, 'user_id': '3', 'movie_id': '1079', 'rating': '5'}
{'_id': 194, 'user_id': '3', 'movie_id': '1615', 'rating': '5'}
{'_id': 196, 'user_id': '3', 'movie_id': '1259', 'rating': '5'}
{'_id': 198, 'user_id': '3', 'movie_id': '2167', 'rating': '5'}
{'_id': 201, 'user_id': '3', 'movie_id': '260', 'rating': '5'}
{'_id': 209, 'user_id': '3', 'movie_id': '1266', 'rating': '5'}
{'_id': 210, 'user_id': '3', 'movie_id': '733', 'rating': '5'}
{'_id': 213, 'user_id': '3', 'movie_id': '2355', 'rating': '5'}
{'_id': 214, 'user_id': '3', 'movie_id': '1197', 'rating': '5'}
{'_id': 215, 'user_id': '3', 'movie_id': '1198', 'rating': '5'}
{'_id': 216, 'user_id': '3', 'movie_id': '1378', 'rating': '5'}
{'_id': 219, 'user_id': '3', 'movie_id': '3552', 'rating': '5'}
{'_id': 220, 'user_id': '3', 'movie_id': '1304', 'rating': '5'}
{'_id': 226, 'user_id': '3', 'movie_id': '3671', 'rating': '5'}
{'_id': 231, 'user_id': '3', 'movie_id': '1136', 'rating': '5'}

15 associated movie titles:
Fish Called Wanda, A
Edge, The
Stand by Me
Blade
Star Wars: Episode IV - A New Hope
Unforgiven
Rock, The
Bug's Life, A
Princess Bride, The
Raiders of the Lost Ark
Young Guns
Caddyshack
Butch Cassidy and the Sundance Kid
Blazing Saddles
Monty Python and the Holy Grail
```

The code example begins by importing sys, os, and custom class conn. Function stages() uses a three-stage aggregation pipeline. The 1st stage finds all ratings of 5 from user 3. The 2nd stage projects the fields to be displayed. The 3rd stage limits the number of documents returned. It is important to include a limit stage, because the results database is big and pipelines have size limitations. Function match_item() uses the aggregation pipeline to match records to criteria. The main block begins by using the stages() pipeline to return all ratings of 5 from user 3. The code continues by iterating this data and using the match_item() pipeline to get the titles that user 3 rated as 5. The pipeline is an efficient method to query documents from MongoDB, but takes practice to get acquainted with its syntax.

Twitter

Twitter is a fantastic source of data because you can get data about almost anything. To access data from Twitter, you need to connect to the Twitter Streaming API. Connection requires four pieces of information from Twitter–API key, API secret, Access token, and Access token secret (encrypted). After you register and get your credentials, you need to install a Twitter API. I chose the Twitter API TwitterSearch, but there are many others.

The 1st code example creates JSON to hold my Twitter credentials (insert your credentials into each variable):

```
import json

if __name__ == '__main__':
    consumer_key = ''
    consumer_secret = ''
    access_token = ''
    access_encrypted = ''
    data = {}
    data['ck'] = consumer_key
    data['cs'] = consumer_secret
    data['at'] = access_token
    data['ae'] = access_encrypted
    json_data = json.dumps(data)
    header = '[\n'
    ender = ']'
    obj = open('data/credentials.json', 'w')
    obj.write(header)
    obj.write(json_data + '\n')
    obj.write(ender)
    obj.close()
```

I chose to save credentials in JSON to hide them from view. The code example imports the json library. The main block saves credentials into JSON.

The 2nd code example streams Twitter data using the TwitterSearch API. To install: *pip install TwitterSearchAPI*.

```python
from TwitterSearch import *
import json, sys

class twitSearch:
    def __init__(self, cred, ls, limit):
        self.cred = cred
        self.ls = ls
        self.limit = limit
    def search(self):
        num = 0
        dt = []
        dic = {}
        try:
            tso = TwitterSearchOrder()
            tso.set_keywords(self.ls)
            tso.set_language('en')
            tso.set_include_entities(False)
            ts = TwitterSearch(
                consumer_key = self.cred[0]['ck'],
                consumer_secret = self.cred[0]['cs'],
                access_token = self.cred[0]['at'],
                access_token_secret = self.cred[0]['ae']
                )
            for tweet in ts.search_tweets_iterable(tso):
                if num <= self.limit:
                    dic['_id'] = num
                    dic['tweeter'] = tweet['user']['screen_name']
                    dic['tweet_text'] = tweet['text']
```

```python
                dt.append(dic)
                dic = {}
            else:
                break
            num += 1
    except TwitterSearchException as e:
        print (e)
    return dt

def get_creds():
    with open('data/credentials.json') as json_data:
        d = json.load(json_data)
        json_data.close()
    return d

def write_json(f, d):
    with open(f, 'w') as fout:
        json.dump(d, fout)

def translate():
    return dict.fromkeys(range(0x10000, sys.maxunicode + 1),
    0xfffd)

def read_json(f):
    with open(f) as f:
        return json.load(f)

if __name__ == '__main__':
    cred = get_creds()
    ls = ['machine', 'learning']
    limit = 10
    obj = twitSearch(cred, ls, limit)
    data = obj.search()
    f = 'data/TwitterSearch.json'
```

```
write_json(f, data)
non_bmp_map = translate()
print ('twitter data:')
for row in data:
    row['tweet_text'] = str(row['tweet_text']).
    translate(non_bmp_map)
    tweet_text = row['tweet_text'][0:50]
    print ('{:<3}{:18s}{}'.format(row['_id'],
    row['tweeter'], tweet_text))
print ('\nverify JSON:')
read_data = read_json(f)
for i, p in enumerate(read_data):
    if i < 3:
        p['tweet_text'] = str(p['tweet_text']).
        translate(non_bmp_map)
        tweet_text = p['tweet_text'][0:50]
        print ('{:<3}{:18s}{}'.format(p['_id'],
        p['tweeter'], tweet_text))
```

Output:

```
twitter data:
0   TradingEnginee4    RT @sonmdevelopment: Great news! Running ML algori
1   lrsevey            #RT @Nexosis: How Businesses Are Using AI and Mach
2   ICM_Change         Digital transformation: How machine learning could
3   RawadKhazem        RT @SASsoftware: The ultimate artificial intellige
4   F5Security         "#CISOs look to #MachineLearning to augment securi
5   jayhinman          RT @Ascendify: Bringing talent and intelligence to
6   meisshaily         Understanding Machine Learning [INFOGRAPHIC] https
7   eriningrassia      RT @DeepLearn007: Machine Learning &amp; Marketing
8   trishia_ani        RT @wef: A computer was asked to predict which sta
9   AmandaMSaunders    RT @dougtraill: Hear from Volkswagen at #GTC18 on
10  DD_Bun_            Machine Learning: What's in It for Business? https

verify JSON:
0   TradingEnginee4    RT @sonmdevelopment: Great news! Running ML algori
1   lrsevey            #RT @Nexosis: How Businesses Are Using AI and Mach
2   ICM_Change         Digital transformation: How machine learning could
```

The code example begins by importing TwitterSearch, json, and sys libraries. Class twitSearch streams Twitter data based on Twitter credentials, a list of keywords, and a limit. Function get_cred() returns Twitter credentials from JSON. Function write_json() writes data to JSON. Function translate() converts streamed data outside the Basic Multilingual Plane (BMP) to a usable format. Emojis, for example, are outside the BMP. Function read_json() reads JSON. The main block begins by getting Twitter credentials, creating a list of search keywords, and a limit. In this case, the list of search keywords holds machine and learning, because I wanted to stream data about machine learning. Limit of ten restricts streamed records to ten tweets. The code continues by writing Twitter data to JSON, translating tweets to control for non-BMP data, and printing the tweet. Finally, the code reads JSON to verify that the tweets were saved properly and prints a few.

Web Scraping

Web scraping is a programmatic approach for extracting information from websites. It focuses on transforming unstructured HTML formatted data into structured data. Web scraping is programmatically intensive because of the unstructured nature of HTML. That is, HTML has few if any structural rules, which means that HTML structural patterns tend to differ from one website to another. So, get ready to write custom code for each Web scraping adventure.

The code example scrapes book information from a popular technical book publishing company. The 1st step is to locate the webpage. The 2nd step is to open a window with the source code. The 3rd step is to traverse the source code to identify the data to scrape. The 4th step is to scrape.

With Google Chrome, click More tools and then Developer tools to open the source code window. Next, hover the mouse cursor over the source until you find the data. Move down the source code tree to find the tags you want to scrape. Finally, scrape the data.

To install 'BeautifulSoup', *pip intall BeautifulSoup.*

```python
from bs4 import BeautifulSoup
import requests, json

def build_title(t):
    t = t.text
    t = t.split()
    ls = []
    for row in t:
        if row != '-':
            ls.append(row)
        elif row == '-':
            break
    return ' '.join(ls)

def release_date(r):
    r = r.text
    r = r.split()
    prefix = r[0] + s + r[1]
    if len(r) == 5:
        date = r[2] + s + r[3] + s + r[4]
    else:
        date = r[2] + s + r[3]
    return prefix, date

def write_json(f, d):
    with open(f, 'w') as fout:
        json.dump(d, fout)

def read_json(f):
    with open(f) as f:
        return json.load(f)
```

```python
if __name__ == '__main__':
    s = ' '
    dic_ls = []
    base_url = "https://ssearch.oreilly.com/?q=data+science"
    soup = BeautifulSoup(requests.get(base_url).text, 'lxml')
    books = soup.find_all('article')
    for i, row in enumerate(books):
        dic = {}
        tag = row.name
        tag_val = row['class']
        title = row.find('p', {'class' : 'title'})
        title = build_title(title)
        url = row.find('a', {'class' : 'learn-more'})
        learn_more = url.get('href')
        author = row.find('p', {'class' : 'note'}).text
        release = row.find('p', {'class' : 'note date2'})
        prefix, date = release_date(release)
        if len(tag_val) == 2:
            publisher = row.find('p', {'class' : 'note
            publisher'}).text
            item = row.find('img', {'class' : 'book'})
            cat = item.get('class')[0]
        else:
            publisher, cat = None, None
            desc = row.find('p', {'class' : 'description'}).
            text.split()
            desc = [row for i, row in enumerate(desc) if i < 7]
            desc = ' '.join(desc) + ' ...'
        dic['title'] = title
        dic['learn_more'] = learn_more
        if author[0:3] != 'Pub':
```

```
            dic['author'] = author
        if publisher is not None:
            dic['publisher'] = publisher
            dic['category'] = cat
        else:
            dic['event'] = desc
        dic['date'] = date
        dic_ls.append(dic)
    f = 'data/scraped.json'
    write_json(f, dic_ls)
    data = read_json(f)
    for i, row in enumerate(data):
        if i < 6:
            print (row['title'])
            if 'author' in row.keys():
                print (row['author'])
            if 'publisher' in row.keys():
                print (row['publisher'])
            if 'category' in row.keys():
                print ('Category:', row['category'])
                print ('Release Date:', row['date'])
            if 'event' in row.keys():
                print ('Event:', row['event'])
                print ('Publish Date:', row['date'])
            print ('Learn more:', row['learn_more'])
            print ()
```

Output:

```
Going Pro in Data Science
By Jerry Overton
Publisher: O'Reilly Media
Category: book
Release Date: March 15, 2016
Learn more: http://www.oreilly.com/data/free/going-pro-in-data-science.csp

2015 Data Science Salary Survey
By John King, Roger Magoulas
Publisher: O'Reilly Media
Category: book
Release Date: September 11, 2015
Learn more: http://www.oreilly.com/data/free/2015-data-science-salary-survey.csp

2016 Data Science Salary Survey
By John King, Roger Magoulas
Publisher: O'Reilly Media
Category: book
Release Date: September 16, 2016
Learn more: http://www.oreilly.com/data/free/2016-data-science-salary-survey.csp

Ten Signs of Data Science Maturity
By Peter Guerra, Kirk Borne
Publisher: O'Reilly Media
Category: book
Release Date: March 14, 2016
Learn more: http://www.oreilly.com/data/free/ten-signs-of-data-science-maturity.csp

Statistics for Data Science
By James Miller
Publisher: Packt Publishing
Category: book
Release Date: November 2017
Learn more: http://shop.oreilly.com/product/9781788290678.do

Scalable Data Science on a Laptop
By Alice Zheng
Event: Hosted By: Ben Lorica Watch the webcast ...
Publish Date: February 12, 2016
Learn more: http://www.oreilly.com/pub/e/3124
```

The code example begins by importing BeautifulSoup, request, and json libraries. Function build_title() builds scraped title data into a string. Function release_date() builds scraped date data into a string. Function write_json() and read_json() write and read JSON respectively. The main block begins by converting the URL page into a BeautifulSoup object. The code continues by placing all article tags into variable books. From exploration, I found that the article tags contained the information I wanted to scrape. Next, each article tag is traversed. Scraping would have been much easier if the information in each article tag was structured consistently. Since it was not, the logic to extract each piece of information is extensive. Each piece of information is placed in a dictionary element, which is subsequently appended to a list. Finally, the list is saved to JSON. The JSON is read and a few records are displayed to verify that all is well.

Index

© David Paper 2018
D. Paper, *Data Science Fundamentals for Python and MongoDB*,
https://doi.org/10.1007/978-1-4842-3597-3

M

N

O

Printed in the United States
By Bookmasters